WITHDRAWN
NDSU

CONCEPTUAL CHANGE

SYNTHESE LIBRARY

MONOGRAPHS ON EPISTEMOLOGY,

LOGIC, METHODOLOGY, PHILOSOPHY OF SCIENCE,

SOCIOLOGY OF SCIENCE AND OF KNOWLEDGE,

AND ON THE MATHEMATICAL METHODS OF

SOCIAL AND BEHAVIORAL SCIENCES

Editors:

DONALD DAVIDSON, *Rockefeller University and Princeton University*

JAAKKO HINTIKKA, *Academy of Finland and Stanford University*

GABRIËL NUCHELMANS, *University of Leyden*

WESLEY C. SALMON, *Indiana University*

CONCEPTUAL CHANGE

Edited by

GLENN PEARCE *and* PATRICK MAYNARD
The University of Western Ontario, London, Ontario, Canada

D. REIDEL PUBLISHING COMPANY

DORDRECHT-HOLLAND / BOSTON-U.S.A.

Library of Congress Catalog Card Number 72-93270

ISBN 90 277 0287 X

Published by D. Reidel Publishing Company,
P.O. Box 17, Dordrecht, Holland

Sold and distributed in the U.S.A., Canada, and Mexico
by D. Reidel Publishing Company, Inc.
306 Dartmouth Street, Boston,
Mass. 02116, U.S.A.

BD
373
C66

All Rights Reserved
Copyright © 1973 by D. Reidel Publishing Company, Dordrecht, Holland
No part of this book may be reproduced in any form, by print, photoprint, microfilm,
or any other means, without written permission from the publisher

Printed in The Netherlands by D. Reidel, Dordrecht

to Julius Weinberg

TABLE OF CONTENTS

INTRODUCTION	IX
KENDALL L. WALTON / Linguistic Relativity	1
PAUL ZIFF / Something About Conceptual Schemes	31
ROMANE CLARK / Prima Facie Generalizations	42
ROBERT W. BINKLEY / Change of Belief or Change of Meaning?	55
WILFRID SELLARS / Conceptual Change	77
KEITH LEHRER / Evidence, Meaning and Conceptual Change: A Subjective Approach	94
STEPHAN KÖRNER / Logic and Conceptual Change	123
JOSEPH S. ULLIAN / Some Comments on Professor Körner's Paper	137
DONALD HOCKNEY / Conceptual Structures	141
NORETTA KOERTGE / Theory Change in Science	167
HILARY PUTNAM / Explanation and Reference	199
ROBERT BARRETT / Referential Indeterminacy: A Response to Professor Putnam	222
N. L. WILSON / On Semantically Relevant Whatsits: A Semantics for Philosophy of Science	233
GENERAL BIBLIOGRAPHY	246

INTRODUCTION

During Hallowe'en of 1970, the Department of Philosophy of the University of Western Ontario held its annual fall colloquium at London, Ontario. The general topic of the sessions that year was conceptual change. The thirteen papers composing this volume stem more or less directly from those meetings; six of them are printed here virtually as delivered, while the remaining seven were subsequently written by invitation.

The programme of the colloquium was to have consisted of major papers delivered by Professors Wilfrid Sellars, Stephan Körner, Paul Ziff and Hilary Putnam, with shorter commentary thereupon by Professors Robert Binkley, Joseph Ullian, Jerry Fodor and Robert Barrett, respectively. And that is the way it happened, with one important exception: at the eleventh hour, Sellars and Binkley exchanged roles. This gave Binkley the rather unusual and challenging task of providing a suitable Sellarsian answer to a question not of his own asking – for Binkley's paper was written under Sellars' original title. Sellars' own contribution to the volume is perhaps more nearly what he would have presented as main speaker than a direct response to Binkley. However, it has seemed best, on balance, to attempt no further stylistic accommodation of the one paper to the other; their mutual philosophical relevance will be evident in any case. The editors would here like to extend special thanks to both Sellars and Binkley for their extraordinary efforts under the circumstances.

Although participants in the colloquium were encouraged to take a free hand in their several approaches to the theme of conceptual change, a number of points of contact have emerged – some anticipated, some happy surprises. Since the final arrangement of the book has been made as much with an eye to these points of contact as to the format of the original programme, a word or two by way of editorial comment is here in order.

Undergoing a change of mind is no uncommon thing; yet not every such change will count as a *conceptual* change. Perhaps the first question one ought here to ask is simply what conceptual changes are. A characterization often enough heard these years is this: a conceptual change is a change in one's conceptual *scheme*, or – if this is different – one's language. More impersonally, one also hears talk of changes in the conceptual scheme (the language) of science. (It is controversial whether one ought to identify conceptual schemes with languages; but it is safe enough to say that conceptual changes *show up* as changes in language.) Underlying the notion of conceptual change, then, are (at least) two basic questions: What is a conceptual scheme? and, What constitutes a difference among conceptual schemes? In 'Linguistic Relativity', Kendall Walton addresses himself to these two questions, attempting to say something about the former by way of the latter; his paper has been placed first primarily because of its concern with fundamentals.

Ziff's overriding concern is to understand understanding. Specifically, he asks how it is that we succeed at communication in spite of the constant presence of a virtually unending set of possible failures. Consider an example: "A cheetah can outrun a man". But what about a cheetah encumbered by a heavy weight? It seems unquestionable that we understand the initial remark and, indeed, agree with it, subsequent thoughts of encumbered cheetahs notwithstanding. Ziff concludes that success in this depends to a large extent upon the adoption of a certain form of 'representation' or 'projection', and that what form this takes depends upon one's conceptual scheme. Whether someone else understands us depends on his conceptual scheme and the extent to which he can appreciate the form of representation employed.

Jerry Fodor was disinclined to publish his commentary upon Ziff's paper; however, Romane Clark's contribution, though self-contained, serves admirably to fill this gap. Clark offers an account of what he calls "*prima facie* generalizations" (taking Ziff's cheetah-sentence and other examples as instances), which he ties to the theory of subjunctive conditionals in an interesting way. He also sketches a general semantic theory – akin to the semantic techniques of the Hintikka-Kripke sort – in terms of which to understand the logic of *prima facie* generalizations, the main lines of which were developed earlier in collaboration with Binkley.

The papers of Binkley and Sellars, already mentioned, deal with the question of whether conceptual change is to be thought of in epistemic or semantic categories – whether, that is to say, what has changed are certain *beliefs* or certain *meanings*. Being ornery, they of course conclude that it is both. The interest of these two papers lies not so much in the conclusions *per se* as in the ground covered getting there.

However we finally describe *what* changes during conceptual change, the mere fact of such change has significant epistemological implications; not the least important are its implications for the notion of evidence. Keith Lehrer attempts to develop a theory of evidence which explains how conceptual change can alter the content of, and inferences from, what we count as evidence.

With Körner's paper we come to a more detailed account of the structure of a conceptual scheme, what Körner calls its "logico-categorial" structure. Taking certain extensions of classical and intuitionistic logic as primary illustrations, he gives special consideration to the sources of pressure toward conceptual change, and to the various ways in which such changes can be accommodated or resisted. Ullian's commentary is in the classic symposiast's mode, focussing mainly on difficulties for Körner. Chief among these is the intelligibility of the notion of alternative *logics* (as opposed, say, to different formal systems) seeming, as it does, to suggest a vantage point outside one's conceptual scheme from which to pass judgment.

The question of logical pluralism comes in for further discussion in the next contribution, albeit with a rather different notion of conceptual structures in view. Indeed, it is the major business of Donald Hockney's paper to develop a theory of conceptual structures which can take account of a variety of types of theoretical difference, and, hence, theoretical change. The development is by way of contrast with views of Quine and, though motivated largely by an interest in physical theories, has applications to logic as well as to natural languages.

The thorny question of how to characterize scientific theories, their criticism and evolution, is continued by Noretta Koertge, who attempts to give an account of scientific change which is a faithful, if idealized, description of actual scientific development as well as a rational analysis of methodological decisions in science – and all this while doing justice to the historical phenomena underscored these days by such writers as Kuhn and Feyerabend.

If Körner's treatment of theoretical change is etiological, Putnam's is morphological. A main objective of the latter's contribution is to sketch a theory of meaning which allows for the apparent facts that (a) concepts which are not strictly true of anything may nevertheless refer to something, and that (b) concepts in different scientific theories may refer to the *same* things. Following Shapere, Putnam calls terms for such concepts *transtheoretical*. The key ingredient in the theory is the thesis that the semantic import of such terms depends primarily upon their reference; that, in turn, is fixed chiefly by the fact that users of these terms are, or are 'causally linked to', individuals in a position to *introduce* them, and not by anyone's knowledge of necessary and sufficient conditions of their application. Development of this idea – which Putnam acknowledges to draw heavily upon Kripke's work on proper names – is applied *inter alia* to physical magnitude terms and natural kind words. Much of the last part of the paper is devoted to criticism of positivistic theories of science which Putnam sees as involving an "idealist" theory of meaning, antithetical to his own position.

Barrett's reply challenges an assumption which he takes to be fundamental to Putnam's entire programme: namely, that the referentiality of the key theoretical terms here in question can be determined at all. Specifically, Barrett argues that Putnam's 'causal' theory is designed only to identify the referents of referring expressions, but in no way helps us to discover *which* expressions are referential in the first place. Into this breach N. L. Wilson steps; in his own unique style, he proposes a semantic theory in which he attempts to retain the centrality of the role of reference (his term is "significance"), but dissociated from the Kripke-Putnam 'causal' theory.

An extensive bibliography of recent works bearing on the subject of conceptual change completes the volume. The editors thank Joseph and Donna Pitt for their patient and thorough work constructing this bibliography on an open-ended topic.

The editors thank O. M. Hitchins and Janice Hall for contributions to the organization of the colloquium, Frances Aird, Linda Palmer, Alice Smith, and Pauline Campbell for preparation of the manuscript. Publication of this volume has been assisted by a subvention from the University of Western Ontario, to which we also express gratitude for support of the colloquium series.

G. PEARCE
P. MAYNARD

KENDALL L. WALTON

LINGUISTIC RELATIVITY

I. INTRODUCTION

The idea that different languages of certain sorts somehow represent, or reflect, or embody, different ways of experiencing, perceiving, or thinking about the world, or different ways of 'ordering' the 'data' of experience – let us say different *'conceptual schemes'* – is an intriguing one. Benjamin Whorf is perhaps its most notable exponent;[1] indeed it is sometimes referred to as the 'Whorf hypothesis.' But its roots go back at least to Kant, and it has been echoed in various forms by a diverse collection of psychologists, linguists, anthropologists, and philosophers.

Unfortunately, however, statements of the Whorf hypothesis and related ideas are for the most part intolerably obscure, mainly because of the obscurity of notions of 'conceptual schemes.' It is far from evident what a way of perceiving, experiencing, or thinking about the world is, or what it is for different people to perceive, exprience, or think about the world in different ways. Yet vague expressions like these are frequently all we are offered in explanation of the 'conceptual schemes' (or whatever they are called) which supposedly are tied up with different languages. My objective in this paper is to begin to clarify *one* sort of thing that might reasonably be meant by 'conceptual scheme.' I will approach this task through consideration of certain features of languages, and so will aim to make sense of the Whorf hypothesis. But I must emphasize that my discussion is not meant to be an explication of the writings of Whorf or of anyone else, although I think it is generally consonant with much that has been said on the topic. Also, there are no doubt many different kinds of things which could plausibly count as conceptual schemes, and what I have to say is by no means intended to cover all of them.

I will concentrate on the notion of *differences* of conceptual scheme. This seems the obvious place to begin in attempting to understand what

Pearce and Maynard (eds.), Conceptual Change, 1–30. All rights reserved.
Copyright © 1973 by D. Reidel Publishing Company, Dordrecht-Holland.

conceptual schemes are. Statements like the following can be taken as (purported) descriptions of our conceptual scheme:

(a) We intuit the objects of experience in the forms of space and time. (Kant)

(b) We think of the universe as a collection of distinct objects and events. (Whorf[2])

(c) Material objects are the basic particulars in our conceptual scheme. (Strawson[3])

But in order to understand them as statements about our conceptual scheme, rather than merely stylistic variants of statements about the world, such as

(a′)   The objects of experience *are* in the forms of space and time.
(b′)   The universe *is* a collection of distinct objects and events.
(c′)   Material objects *are* the basic particulars.

we would expect to have *some* understanding of what it would be to intuit the objects of experience in forms other than space and time (or in no forms at all), to think of the universe otherwise than as a collection of distinct objects and events, to have a conceptual scheme in which material objects are not the basic particulars. We may or may not think it possible for us, or for anyone or any thing, to have conceptual schemes of these alternative sorts. But explaining what it would *mean* to say that there is a difference of conceptual scheme between us and other beings would seem to be the job to tackle first in trying to make sense of the notion of conceptual schemes, and of descriptions of particular conceptual schemes.

Our starting point is the vague characterization of different conceptual schemes as different *ways* of perceiving or thinking about things (phenomena, events), or different ways of 'ordering' the 'data' of experience. A difference of *concepts* is not sufficient, I take it, for a difference of conceptual scheme. An isolated tribe living in a hot climate may have no concept of snow, but they do not thereby have a different conceptual scheme from ours. Nor do people who lack concepts that we have, not because their environment is different, but because they are physiologically unable to perceive certain features of their surroundings that we can perceive. A tribe of people who are color blind, or who have no perceptual organs sensitive to differences of temperature, may lack our color and temperature concepts. But I will not on this account consider them to have a conceptual scheme different from ours.[4] We and they do not experience phenomena in different ways, but merely experience different phenomena. The difference is in *what* is experienced, rather than in *how* it is experienced.

The clearest cases in which different people can be said to have different ways of ordering data of experience are likely to be ones in which the data that are ordered are the same for both. There will also be cases in which the data are not identical, but are similar, or corresponding, facts about different objects. Let us say that two facts 'correspond' if each is the fact that an object has a certain property, and the property is the same in both cases. Giving sense to the notion of experiencing *non-*corresponding data in different ways seems an especially perplexing task. The non-correspondence of different people's data may by itself account for differences in their experiences, making it unnecessary to postulate difference in *how* they experience their respective data. It also seems to me that different ways of experiencing corresponding data are likely to be more interesting than different ways of experiencing non-corresponding data, even if the latter does make sense.

In any case, I propose to restrict the notion of differences of conceptual scheme to cases in which people have corresponding data. People with different conceptual schemes of the kind I am explicating are (in some sense) at bottom concerned with the same properties of objects; the 'given' is of the same sort for both, but it is somehow 'organized', 'ordered', 'interpreted', differently.

My first objective will be to clarify what might reasonably count as differences of conceptual scheme of this sort. I will sketch simple examples of (possible) societies with languages different from English, which might plausibly be said to have conceptual schemes different from ours.

We must face up to a difficulty which may seem to doom our program, and perhaps the Whorf hypothesis, from the start. How could we determine that a language different from ours involves a different conceptual scheme? It seems that we could not hope to determine this about a language which cannot in some way be translated into English. Imagine discovering a tribe of people whose language is unintelligible to us.

No matter how much we try we cannot make sense of the noises they utter; we cannot say in what circumstances a certain sort of noise would be appropriate, or find anything like an English equivalent for it. (We might question whether the noises actually are expressions of a language, but let us suppose that they are.) The incomprehensibility of the tribe's language might be taken as proof that we and they have different concepts. But we would not be entitled to say that their language embodies a different conceptual scheme. So far as we could tell, it merely functions to describe kinds of facts which we are completely unaware of. Speakers of the language might simply have a form of sense perception, a sixth sense, which we lack, and they may lack our senses. So the difference between their language and ours may reflect a difference of data rather than anything which could reasonably count as different ways of ordering or experiencing data.

It may be urged that though we must have some *understanding* of another language in order to know that it involves a different conceptual scheme from ours, we would not have to be able to *translate* or explain the meanings of its expressions in English. I suppose that we could learn a tribe's language as a native child would, simply by living with the people and picking it up, without in any way correlating or associating particular native expressions with particular English ones. But we still could not regard the language as involving a different scheme. We would understand both languages; we would know when and how to use the expressions of English, and when and how to use those of the other language. But we would not see any *connection* between the two languages, and so would have no reason to suppose that they represent different ways of ordering *corresponding* facts. Again we would not have ruled out the possibility that the other language simply serves to talk about facts which are not expressible in English, ones which we became aware of only when we learned the other language. So if we are to find an example of a language different from ours which embodies a different conceptual scheme, and understand it as such, it seems that we must look among possible languages which admit of some sort of translation into English.

But how can a language which *is* translatable into English involve a different conceptual scheme? If we can construct an English sentence synonymous with any sentence of another language it would seem that we can express any idea that its speakers can, and have every concept they have. This appears to leave no room for a difference of conceptual scheme. A mere difference in what sounds or marks are used is obviously not enough. To adopt a code, e.g., Morse Code, in which every statement has an exact English equivalent certainly would not be to adopt a new conceptual scheme. There is such a thing as coming to think in a code, i.e., coming to use it without thinking of the equivalent English expressions. But this is not learning to think differently or experience the world differently in anything like the way people with different conceptual schemes are supposed to; it is just becoming facile with a new notation. Any language which is translatable into English may appear to be in the same boat with Morse Code English. The rules of translation might of course be much more complicated. But so long as translation is possible the difference between another language and English would seem to be *merely* linguistic, just a difference of notation, however complicated it is. We and the speakers of such a language would simply have different ways of saying the same things, different means for communicating the same information. And this is a far cry from having different conceptual schemes.

Thus, we appear to be in a dilemma: If a language can be translated into English it cannot embody a conceptual scheme different from ours. And if it cannot be translated, we cannot reasonably believe that it embodies such a scheme, even if it does in fact. A language embodying a different scheme must be an untranslatable one, but because of its untranslatability we could not be aware of the difference of conceptual scheme.[5] So the Whorfian view that linguistic differences are tied up with conceptual differences of some sort appears to be essentially unverifiable; it is not even theoretically possible, apparently, to find an example which bears it out.[6]

This dilemma is not fatal, as I will demonstrate.

II

The linguistic differences I will be concerned with amount to differences in how languages classify or group things.[7] Consider a society, A, with a language (A-language) which incorporates a different way of classifying birds from ours. A-language contains species names, like our 'swallow', 'sandpiper', 'finch', etc., but with different ranges of application. For instance, there is a species name, 'Ø', in *A*-language which applies to some but not all of the birds we call 'swallows' plus some but not all of the birds we call 'swifts'. Also, *A*'s use relational predicates which function like our "same species" and "different species", but which are of course not true of the same pairs of birds. I assume that we can specify the classes of birds that constitute *A*'s species in English, in terms of birds' colors, shapes, flight patterns, calls, etc., i.e., we can *construct* complicated predicates extensionally equivalent to *A*'s species names. But the crucial difference is that these English predicates contain other predicates (e.g., 'has a brown body', 'flies erratically') whereas *A*'s species names, like ours, do not. In order to mark *A*'s species classifications in English we must refer to other categories of birds. But this is not necessary in *A*-language. I assume that the opposite is true of our species classifications. We might say that English and *A*-language each treats its own species as fundamental categories and the foreign species as derivative ones. I shall argue that we can expect a corresponding difference in what categories are fundamental to our, and to *A*'s, thought and perception, and that this difference qualifies, the dilemma notwithstanding, as a difference of conceptual scheme.

I have assumed that species names in both English and *A*-language are co-extensive with complex predicates involving birds' colors, shapes, etc. Such features are what make swallows swallows, for example, and they can serve as criteria for recognizing them.[8] Of course different swallows may be swallows in virtue of different combinations of features. Let

(1) (x) (x is a swallow if and only if x has features $a, b, c, \ldots$; or $d, e, f, \ldots$; or $g, h, i, \ldots$; or $\ldots$)

be a true statement specifying these combinations, and so providing a rule for determining which birds are swallows. (1) is to be construed as supporting counterfactuals. And I am assuming that it is finitely long.[9] (1) does not allow for combinations of features that constitute borderline cases of swallows; I am ignoring this complexity.

It will be convenient to call the property of being a swallow, and other properties which are similarly based on criteria, *secondary characteristics*, and properties which serve as criteria for secondary characteristics *primary characteristics with respect to those secondary ones*, or just *features*. It should be noted that characteristics which are primary with respect to certain secondary ones may be based on criteria and hence secondary themselves. Having a hooked bill is a (one of many) primary characteristic with respect to being a hawk, but it can be regarded as based on the various details of bills' shapes which make them hooked. If there are properties which are not based on other ones at all, which are not secondary, I will call them *basic characteristics*. I will also speak correspondingly of primary, secondary, and basic *predicates*.

I have been deliberately inexplicit so far concerning the nature of the relation between primary and secondary characteristics. Just how are the latter "based on" the former? One view, which I will call the *reductionist* view, is that secondary characteristics are "reducible" to or "analyzable" in terms of primary ones, that species names, e.g., mean the same as complex primary predicates and hence that statements like (1) are analytic. This is a tempting view. A bird can be described *completely*, it seems, by pointing out its features. Its swallowness, or hawkness, is not an additional property over and above its features; to be a swallow is just to have certain features – some one of the combinations of features specified by (1). On this view secondary predicates function as abbreviations for complex primary ones. To call a bird a swallow is simply an easy way of saying something complicated about its shape, color, size, etc.

When we regard secondary characteristics in this way we are thinking of our species categories as essentially *arbitrary* or *conventional*. Our classifications of birds are not forced on us by the facts of nature; we just happen to have conventional rules for classifying them as we do. We can easily imagine using different rules and coming out with different groupings, such as A's. To adopt a classification like A's would seem to be merely to substitute one set of abbreviations for another. A's often appear to disagree with us about whether two birds are of the same or different types. But (one might think) it does not really make any difference which is said, so long as everyone agrees on the "facts" – i.e., the specific respects in which birds resemble and differ from each other. These facts are the *only* ones involved; and if there is no dispute about them the question of whether to class the two birds together can be decided by simple fiat.

On the reductionist view statements about species in both English and A-language are equivalent to statement about features, which (I am assuming) can be described in both languages; hence the languages are intertranslatable. If so the first horn of the dilemma poses no problem; the danger lies rather with the second horn. It may appear that the two languages merely incorporate different systems of abbreviations for statements common to both, that this is just a difference of arbitrary notation, and hence that there is no better reason for attributing a conceptual scheme different from ours to A's, than to speakers of Morse Code English.

It will be illuminating to contrast tribe A with a tribe that suffers from the opposite problem. Let B's be people who seem to classify things according to tastes differently from how we do. We distinguish between things which taste sweet, things which taste bitter, things which taste salty, etc. (These classes, unlike species categories, have considerable overlap.) B's have similar but non-equivalent groupings. They say that certain things (e.g., chocolate ice cream, garlic, distilled water, a lemon) each has a certain kind of taste, T, which other things (e.g., another lemon) lack. We recognize no (kind of) taste which is shared by the first group of things and not others, and they recognize no taste common to, e.g., the things we call salty tasting and only to them.

Tastes are plausible candidates for *basic* characteristics. (But cf. Section VIII below). It seems that something does not have a salty taste in virtue of any other facts about it, in the way that a bird is a swallow in virtue of its features. There are no rules or criteria for deciding whether something tastes salty; we just taste it and see. If it has a salty taste that is just a matter of fact; not a matter of how we happen to describe or look at any facts. No arbitrary convention is involved. Since we do not use rules to judge tastes, we cannot imagine changing the rules and coming out with different groupings of things, as we can imagine changing the rules for dividing birds into species.

If we met one member of tribe B alone we would probably say that his judgments about what things have similar tastes are just wrong when they conflict with ours, that he has a deficient palate. But on discovering that what he says is consistently confirmed by the rest of his tribe, rather than insist that he and his fellows are all mistaken we would probably hold that they do not mean what we mean by 'same (kind of) taste', and 'different (kind of) taste', that in using these expressions we and they are classifying things according to different properties. We might look for another way of translating their expressions into English. But if we do not succeed a natural conclusion would be that they perceive what they call "tastes" by some sixth sense (or a special sort of taste sense) which we lack, and that their taste properties are properties we simply are not aware of.[10]

Our differences with B's would be puzzling to us, in a way that our differences with A's would not be. We cannot give rules for using B's taste language, as we can for using A's species terms, or in any way translate or paraphrase it in English. Because of this we cannot construe B's as having a different conceptual scheme from ours. So far as we can tell B's just perceive different sorts of properties; they do not experience corresponding facts in different ways.

Thus, whereas the second horn of the dilemma may seem to disqualify A-language as a language which we can understand as involving a conceptual scheme different from ours, B-language is impaled on the first horn. The (apparent) problem with A-language is that it is translatable into English, and the problem with B-language is that it is not.

III

The reductionist view of secondary characteristics seems to me very misleading, at best; secondary characteristics do not differ from basic ones nearly as sharply as that view would suggest. My reasons for saying this will provide the key for a solution of the dilemma that is plaguing us.

Although there are rules or criteria for ascertaining a bird's species on the basis of its features, often we do not use these rules. If we are familiar enough with swallows, for example, we may recognize them without even noticing their features.[11] We may identify a swallow *just* by *looking* at it, in very much the way we tell whether something has a salty taste, without making an inference on the basis of its features. I grant that even in such cases it is the swallow's features that enable us to recognize it; if it had different ones it would not look to us like a swallow. But in recognizing it we need not be at all consicous of its features. Indeed, one might pick out swallows without ever realizing what features are characteristic of them. One might be entirely ignorant of (1), and have no idea how to identify swallows from their features. And even if one does know (1) he may not realize which of the disjuncts of (1) applies to a given swallow when he recognizes it, which of the various different swallow-making combinations of features the bird has. Thus, one does not need to be able to define or analyze species names in terms of birds' features, or in terms of anything else, in order to understand and apply them correctly;[12] they may function exactly like basic predicates.

This may be more evident if we compare other cases. We usually do not recognize our friends by noting their facial and bodily characteristics and applying a rule of inference, though it is these characteristics which make it possible for us to recognize them. We generally can say very little about their features, certainly not enough to enable someone else to recognize them with much confidence, though we have no difficulty recognizing them ourselves. Similarly, one may identify a person's facial expression as one of determination, or chagrin, or despair, without having any idea what it is about the features of his face that makes it that kind of an expression. We no not *infer* that it is a determined expression, but just look and see that it is.[13]

I will assume that A's species names frequently function like basic predicates also. A's can and often do divide birds into their species categories without noticing or considering their features.

The above considerations seem to me to cast considerable doubt on the reductionist view that species names are abbreviations for complex primary predicates (though I shall not attempt to settle the issue here); insofar as we use species names without thinking of birds' features we simply are not treating them as abbreviations. If the reductionist view is dubious the translatability of A-language into English is also, since the species names of A-language were supposed to be translatable in terms of English primary predicates. I must now show that even if we do hold A-language to be untranslatable we can plausibly regard it as escaping the first horn of the dilemma. That horn, which asserts that no untranslatable language can be known to involve a different conceptual scheme, must, to be acceptable, be weakened or construed so as not to exclude A-language. Recall that the point of the first horn was to ensure that people counted as having different conceptual schemes are somehow concerned with the same or corresponding fundamental data, that they are not just aware of different kinds of facts.

As was mentioned, our ability to identify swallows depends on their features, even when we do not take note of the features. But it is not sufficient merely that they *have* the requisite features. In order for us to recognize a swallow just by looking at it, conditions must be favorable for observing its features; conditions must be such that we *could* tell what the bird's features are if that were what we were interested in. We might mistake it for some other kind of bird, if, for instance, it is under ultraviolet light, or at an angle which obscures the shapes of its wings. (I think this amounts to saying that its features function as what gestalt psychologists call "perceptual cues".) This makes it not unreasonable to hold that when we identify swallows just by looking at them, we perceive or experience their features, though *one* important criterion for saying this – our being able to describe the features – is missing. (Compare: we might say that a baseball player who hit a curve ball for a home run *must* have seen the ball curve, even if the thought that it was curving did not go through his mind and he does not recall afterwards that it was a curve ball – unless we think he was just lucky.) I will say that the features (or the facts that the birds possess them) are "given", or are "data" of our experience.

Moreover, we can reasonably regard a bird's features as not just some but *all* of our data relevant to our recognizing it as a swallow, even when we recognize it non-inferentially. The justification for this is the fact (which I assume) that our having those data is not only necessary but (causally) sufficient for our being able to identify the bird as a swallow, given our normal state of mind and nervous system. Our recognition of the swallow does not require any special state of any of our sense organs, or their being receptive to any stimuli, other than what would be necessary for us to recognize the features. Thus there is a sense in which our experience is based *entirely* on these data. Our perceiving that the bird is a swallow can be regarded as simply a way of experiencing its features, rather than an experience of some other datum.

I will assume that for similar reasons when A's apply their species names non-inferentially, birds' features constitute all of the relevant data of their experience. (That this is so could be ascertained experimentally by the ordinary inductive procedures one would use in any case to determine what "cues" are involved in perceptual judgments.) Hence A's are not simply concerned with properties of birds which we do not recognize. We need not suppose that they have anything like a sixth sense. The features which serve as their data are ones which we sometimes identify and talk about explicitly, and which also underlie our own species categories.

So whether or not we regard A-language as translatable into English, the dilemma's first horn is avoided. We can at least *explain* A-language in English; it is no mystery to us. We can give rules in terms of features for applying the secondary predicates of A-language. And, although A's do not always use these rules in applying their predicates, we can understand and explain how they do it. We know what it is for certain observable but unnoticed facts to enable one to ascertain without inference some other fact; and we know what kinds of facts in this way enable A's to apply their secondary predicates, even though we cannot apply them this way. Because we can thus explain the workings of A-language we can see that our and A's experience is based on corresponding "data". So this sort of explanation is all that can legitimately be required by the first horn of the dilemma.[14]

IV

It remains to be shown that A-language escapes the second horn, that we and A's can reasonably be said to experience the corresponding data in different ways. (This will require making one further assumption about A's.) The crucial difference between us and A's is a difference in which classes of objects are "natural" (rather than "arbitrary") for us and which are "natural" for them, i.e., a difference in what we and they recognize or regard as "simple" properties. A class is "natural", let us say, just in case there is a (single) respect in which all members of the class are alike and are different from everything else. This means that there will be a *property* possessed by every member of a given natural class and by nothing else. But not just any property will do, unless the notion of natural classes is to be utterly trivial. Every group of things whatever shares at least a *disjunctive* common and peculiar property. Beethoven's piano, the number 17, the Roman Empire, and Abraham Lincoln, for example, share the property of being *either* Beethoven's piano or the number 17 or the Roman Empire or Abraham Lincoln, and that is a property possessed by no other thing. But we do not seriously look on this disjunctive property as a respect in which those four entities are alike or resemble each other, something which makes it non-arbitrary to class them together. The same is true generally of properties that strike us as disjunctive, including ones which pick out "open" classes. We would think it sophistry, or a joke, to say that all pianos, empires, numbers and men are alike in that each of them is either *a* piano, *an* empire, *a* number, or *a* man, or that this property makes it non-arbitrary to class all of these things together. *Different* properties occur in these objects – some are pianos, other empires, others numbers, and still others men. And since these properties are different they do not constitute a respect of similarity. The common disjunctive property does not prevent the class from being arbitrary, for it inevitably invites the question: why put just *these* disjuncts together to define a single class?, a question which does not arise with what seem to be simple non-disjunctive properties.[15]

Arbitrary (i.e., non-natural) classes are not all *equally* arbitrary. "Family resemblance" classes, classes which are united by disjunctive properties such that the disjuncts are conjunctions with many conjuncts in common, are perhaps less arbitrary than others. But there still seems to be a distinction between disjunctive properties of the family resemblance variety and simple properties, and a corresponding distinction between family resemblance and (completely) natural classes. Various subsets of a family resemblance class have various (simple) properties in common, but this does not make all members of the class alike in a *single* respect.

I do not assume that there is an "objective" or "absolute" difference between natural and arbitrary classes, or between simple and disjunctive properties. But the difference is clearly not just a linguistic, language relative one either. Our intuitions about which properties are simple and which disjunctive are not explained merely by the fact that some *predicates* in our language are grammatically simple and others are grammatically disjunctive, nor even by that plus the fact (if it is one) that only things which share a property expressed by a grammatically simple predicate can be called in English "similar in a single respect". We could deliberately introduce into English a simple predicate, '*penm*', true of all and only pianos, empires, numbers, and men. But this change in our language does not make us any more inclined to regard pianos, empires, numbers, and men as constituting a natural class, and the property of being a *penm* seems no less disjunctive than the property of being either a piano, an empire, a number, or a man did previously.

Thus, it seems, properties can be "simple" and classes "natural" in senses in which this is independent of any *language*. But we are not forced to recognize absolute senses of those expressions; there is a middle ground. I will explicate *society relative* notions of simple properties and natural classes. The property of being *penm* would not be simple and the class of *penm* things would not be natural *for us*, i.e., relative to the English speaking community, even if the word '*penm*' were added to the English language. I leave open the questions of what sense, if any, can be given to the idea of a property being simple or a class natural *simpliciter*, and of what properties and classes are such. The society relative notions are all that is required for our discussion of conceptual schemes.

Consider a tribe of people, C_1, whose species names are coextensive with ours, but who apply them without *ever* making inferences on the basis of birds' features. Imagine further that for some reason they have never learned to recognize the relevant features, and that their language contains no means of describing them. They can identify swallows and sandpipers on sight, but have no notion of the features that make swallows swallows and sandpipers sandpipers. Consequently, they cannot *explain* how to use the expressions "swallow" and "sandpiper", just as we cannot explain how to use "salty taste".

C_1's would not construe bird species as based on other characteristics, in the way that we do. They would not even be tempted to consider species names as abbreviations for complex primary predicates. C_1 philosophers probably would take a rather Platonic view of their species. They would not regard it as at all arbitrary or a matter of decision which birds to group together, any more than we so regard our groupings of things according to taste. They could not justify classing certain birds together by citing a conventional rule, but only by asserting that those birds share a certain property (e.g., swallowness), that they resemble each other in a certain respect. One bird just *is*, or is not, of the same kind as another, as far as they are concerned; this is simply a fact, and has nothing to do with anyone's language or way of looking at things. A C_1 could not handle what we would call a borderline case in the way we might. He may confess that he cannot tell whether a certain bird is a swallow or a swift, and not mean by this merely that he does not know which to *call* the bird. He would not resort to saying that it does not matter which we call it so long as we agree on the "facts". For the "facts", the specific respects in which the bird resembles and differs from unambiguous cases of swallows and swifts, are ones which he does not recognize. There is no escape for him from the question of what species the bird is in, even if he cannot answer that question.

C_1's also would lack our tolerance for different classifications of birds. Let C_2 be a tribe of people with species coextensive with A's, i.e. different from ours and C_1's, but who, like C_1's, have no notion of the features underlying their species and never use criteria for determining birds' species. A confrontation between C_1 and C_2 would produce the kind of puzzlement we and B's would occasion in each other. The C tribes could not resolve their differences as to what birds belong in the same species in the way that we and A's can. Neither group can devise rules for applying the other's species names. C_1's simply could not understand what C_2's species names mean, and would probably think C_2's are grouping birds according to properties they are blind to. They would not say that the different classification systems constitute different ways of ordering corresponding facts, but rather that the two tribes are simply aware of different kinds of facts. Because of the first horn of the dilemma, then, C_1's could not know that C_2's have a different conceptual scheme (though *we* might hold this), just as we cannot say that B's have a different one.

But C_1 and C_2 *clearly* escape the second horn, as do we and B. Obviously it cannot be argued that the languages of C_1 and C_2 do not embody different conceptual schemes on the ground that they are intertranslatable. The two tribes do not merely employ different terminologies. The difference between them is obviously deeper than that. They exhibit the mutual mystification which, the second horn implies, tribes with different conceptual schemes necessarily have toward each other. Each tribe regards birds as naturally falling into different groups, and finds the other's groupings incomprehensible.

It seems reasonable to consider the difference between C_1 and C_2 a difference in what classes are natural for each. Being a swallow is a simple property for C_1's, a respect in which all swallows are alike and different from everything else. But swallowness is not a simple property for C_2's; it is not a property that C_2's recognize at all.[16] And assuming that no other property which is simple for them happens to be true of all and only swallows, the class of swallows is arbitrary for them. C_2's species are similarly natural classes for them and arbitrary classes for C_1. But more needs to be said about the conditions in which, in general, a class will be natural, or a property simple, for a group of people.

What should be said about a society of people who identify swallows and sandpipers sometimes by inferring from their features with the help of disjunctive rules, and sometimes just by looking at them, or a society of people some of whom identify species in one way and others in the other way?[17] The answer hinges on which method of identification is considered to have priority. If the people trust rule-applying identifications only because, and only to the extent that, they are expected to coincide with non-rule-applying identifications, i.e., if they would accept the result of non-rule-applying identifications should the two tests conflict, swallowness will count as a simple property for them (and the class of swallows a natural class.) But if the rule-applying test takes precedence over the non-rule-applying one in this way, swallowness will be a disjunctive property for them.[18]

Let us assume that A's give clear priority to recognizing members of their species just by looking at them, although they do sometimes recognize them by using a rule; their species are natural classes for them. A's treat their secondary predicates as basic ones. When they recognize birds just by looking they are thinking solely in terms of their species categories; the only questions they are concerned with are what birds belong to what species. And, even when they take a bird's features as evidence of its species, in stating what species it is in they are not talking about its features any more that C_2's are (just as in asserting that Jones is guilty of robbing a bank is not saying anything about his fingerprints, even if Jones' fingerprints are the evidence for his guilt). They regard their species not as conventional or based on arbitrary rules, but as rooted in the nature of things. ϕ's strike them as falling *naturally* into a single class, as sharing a single respect of resemblance. ϕ's just appear to them to be birds of a single kind.

A's species are not natural classes for us (assuming that no properties which are simple for us happen to coincide with their species). We cannot help but be quite conscious of the arbitrariness of A's classifications, for we *cannot* categorize birds as they do without having in mind the arbitrary rules for doing so. We cannot think of the class of ϕ's except as united by a complex disjunctive property.

Whether our species are natural classes for us has no straight-forward answer, though they clearly are not natural for A's. Probably ornithologists give priority to rule-applying identifications of birds, but many others of us probably do not, and most of us are uncertain or switch back and forth depending on the context. Our species are natural for us to whatever extent non-rule-applying recognition has priority.

But the fact that certain classes are natural for A's but not for us is all we need consider. Given that we and A's nevertheless have corresponding data this difference may be considered a difference of conceptual scheme, a difference in ways of experiencing or ordering the data.

Different classifications which simply reflect the way things are – as groupings of objects according to taste seem to – classifications which are in no way arbitrary or conventional, would merely concern different sorts of facts. On the other hand, different classifications which are straightforwardly arbitrary, mere linguistic devices to facilitate the exchange of information about the facts, would just constitute different ways of talking about the facts. In either case no difference of conceptual scheme would be involved. But classifications may fall between these extremes, and one can see that they do. When we and A's stop to consider the features which *sometimes* serve as criteria for our respective classifications of birds, we can see that there is an arbitrary element in them, that different systems of classification would fit the *data* equally well. But in the normal course of events it is easy to lose sight of these criteria and treat bird classifications as we treat taste groupings. Insofar as priority is given to judgments made in this spirit the classifications lose their arbitrariness as far as we are concerned.

A's species classifications, and to some extent ours, are thus neither mere reflections of the facts, nor transparent patterns superimposed on them for convenience of communication; they are comparable neither to the outlines of continents on a map, nor to the arbitrary lines of latitude and longitude. They are better construed as forms or concepts through or by means of which we and A's experience the facts. They – the classifications themselves, not merely the visual and auditory symbols which mark them – are part of the mechanism of our and A's thought and perception. They embody two alternative ways of ordering the data of experience.

V

The notion of differences of conceptual scheme illustrated above can be extended, with relatively minor modifications, to a much broader range of cases than this first simple example may lead one to suppose. The key to this notion is the possibility of applying secondary predicates without using rules relating them to the primary characteristics on which they are based. A *sufficient* condition for applying them this way is applying them without being aware of the primary characteristics. But this is not a necessary condition. One may not be using a rule in recognizing a swallow even if he does notice its features, if his awareness of the features plays no part in his recognition of the swallow. There are other cases, however, in which recognition of a secondary characteristic does depend on awareness of features, but yet we do not in any straightforward sense use a rule of inference.

An activity is, or is not, a game because of certain of its other characteristics: whether it is done for fun, or for some practical reason, whether it is competitive, whether it involves winning or losing, etc. So I will call these primary characteristics, or features, with respect to the secondary characteristic of being a game. Again, any of various different sets of features may make an activity a game. Let

(2) (x) (x is a game if and only if x has features $a, b, c, \ldots$; or $d, e, f, \ldots$; or $g, h, i, \ldots$; or $\ldots$)

be a true statement, which a foreigner might memorize and use as a rule for deciding which activities to call "games".

The features which make activities games are not often observable at a glance, and so are unlikely to be combined in a single visual gestalt by which we identify games. Identifying games normally requires our knowing their features. This is especially true when we identify them from (written or verbal) descriptions of their features; we must understand the descriptions and hence must know the features. But identifying games is usually, nevertheless, importantly like recognizing swallows just by looking at them. We need not recall previously learned rules and subsume each particular case under them, as the foreigner may. We likely have never tried to formulate such rules, or even wondered which features are relevant. When we learn what features an activity has it just *seems* to us to be a game, or not one; we are simply inclined to call it one or the other. Perhaps we do judge which it is from, or on the basis of, its features. But we do not employ a rule of inference in doing so.

It will be objected that we could not identify games from their features without *knowing* ("in some sense") what the relevant rule is (i.e., (2)), and that the necessity of this knowledge makes it plausible to hold that we *use* the rule. I contend that the sense in which we must know the rule is marginal at best, and does not support the claim that we use it.

The fact that we do not have the rule in mind when we identify games and have never formulated it does not of course show that we do not know it. I know the spelling of a word, for example, even if I am not thinking about its spelling and never have, if, were I asked, I could spell it correctly. But clearly we could not even recite (2) on demand. We could hardly begin to specify the criteria for being a game if someone asked us to, though we are perfectly able to use 'game' in the right cases.

Perhaps knowing the rule requires merely our being able to *recognize* it if it is formulated for us. But can we even do this? The whole of (2) would probably confuse us by its complexity. So imagine that we are asked one at a time about more specific rules derived from (2), e.g.,

(2') (x) (If x has features $a, b, c, ..., x$ is a game.)

If we know that an activity has features $a, b, c, ...$ we will identify it as a game. But would we realize, if asked, that these features are *sufficient* for making it a game? We might judge an activity to be a game on the basis of a set of features which are not sufficient, e.g., $a, c, ...$. In this case there is another feature, non-b, which, if it belongs to the activity, would prevent it from being a game, and would make us reverse our judgment if we knew about it. In judging the activity to be a game we may not realize that there could be a defeating feature of this sort, unless this is pointed out to us. (Our judgment may nevertheless be justified if, given the kinds of activities that are common, it is unlikely that non-b is present.) Now if we know than an activity has features $a, b, c, ...$ (which are in fact sufficient), we will call it a game, but we may reasonably be unwilling to say that they are sufficient, allowing for the possibility of a defeating condition. So we still need not recognize the rule that *any* activity with these features would be a game. In short, if we know of any activity that it has $a, b, c, ...$ we will know that it is a game, but we do not necessarily know that if any activity has (or had) those features it is (would be) a game.[19]

Furthermore, after identifying a game from, e.g., a verbal description, we may not realize which of its features make it a game. Some of the described features may be irrelevant, and the activity may strike us as a game without our being able to specify which are relevant. (Compare: On reading a novel we may realize that a character is unstable, or pompous, without being able to say which of his actions or remarks gave us this impression. Yet the actions or remarks responsible must be ones we knew about; they could not have had their effect on us if we did not understand the descriptions of them in the novel.) Hence we may be unable to say just what the rule is. In order to identify an activity with $a, b, c,...$ as a game we need not realize that these features are just sufficient, that none is unnecessary, i.e., that the rule involved is (2'), rather than some stronger one.

The main support for the view that we must know the rule for 'game' lies in the fact that we could (presumably) say what it is if we undertook an extensive and laborious investigation, asking ourselves dozens, or hundreds, of questions of the form: "Would an activity with... features be a game?" and piecing together a statement of necessary and sufficient conditions from the answers. (This is roughly how philosophers from Plato on have proceeded in giving "analyses" of concepts.) But this kind of investigation seems to me to have more the character of *discovering* rules than of demonstrating (dispositional) knowledge that we already have. Success is notoriously elusive, as the history of philosophy testifies. Time and again careful formulations of such rules have fallen before counterexamples (which often are excruciatingly evident once they are pointed out.) Our "knowledge" of the rule for 'game', then, is so *highly* dispositional that it seems misleading to call it knowledge.

But suppose that we do call it knowledge. For it to be reasonable to claim that we *use* the rule in deciding what activities are games it should be possible, one would think, to discover that we know it independently of the use we make of it. But the *only* way to discover this is by our decisions as to what sorts of activities to call games. Our "knowing" the rule consists *merely* in our acting *in accordance* with it, i.e., making the right decisions. So how can we be consulting it in making these decisions? Moreover, we cannot appeal to a rule to defend a judgment that an activity is a game (as we might cite a rule book to show that a chess move is proper), for there is no way to settle what the rule is except by considering what we already know to say in such particular cases. Thus, identifying games from descriptions in the way we normally do is like recognizing swallows on sight, in that we are not making use of a rule in either case.[20] When we learn of an activity's features it just *strikes* us as a game (if it is one), as resembling other activities we call games. We are not treating 'game' as an abbreviation for a complex disjunctive predicate about activities' features, for we do not have in mind the relevant disjunctive property.

Moreover, since we do not and cannot identify games by applying rules, non-rule-applying identifications have priority by default. So the property of being a game is simple for us, and the class of games is natural for us.

A person who identifies games only by using rules, or who at least gives priority to identifying them in this way, would regard games very differently. He would be fully aware of the vast variety of combinations of features different games have, and unaware of any respect in which they are all alike. He would see only a *conventional* justification for lumping tennis, chess, professional baseball, solitaire, poker (for money), etc., together, while excluding, e.g., mountain climbing, war, making love, playing the stock market, and playing string quartets.

We must not forget that the property of being a game is based on features of activities. There can be no objection to saying that when we recognize games the features constitute our "data", for we are usually quite conscious of them and explicitly use them as grounds.

It is obvious how to set up another example of a tribe with a different conceptual scheme from ours. Let D's be people who recognize the same features of activities that we do, i.e., whose data corresponds to ours, but for whom different classes of activities based on such features are natural.

VI

We should have a somewhat more precise specification of the minimum conditions in which, in general, two groups of people will be said to have different conceptual schemes of the kind I am considering. First, a preliminary definition: I will say that a property of an object is *apparent* to a person if and only if either (i) the person is in a position to observe that the object has that property (as we are in a position to observe the features of a bird when we tell just by looking that it is a swallow), or (ii) he knows that the object has that property. Two groups of people have different conceptual schemes (of the kind under discussion) if and only if the following conditions are met:

(a) There is (at least) one predicate, P, such that the members of one of the groups, X, give priority to non-rule-applying tests of whether something is P.

(b) There is no predicate, P', extensionally equivalent to P, such that the members of the other group, Y, give priority to non-rule-applying tests of whether something is P'.[21]

(c) P is extensionally equivalent to a disjunctive predicate each of whose disjuncts is either one of the predicates $Q_1 ... Q_n$ (which may or may not exist in an actual language), or a conjunction of them, and none of the disjuncts entails any other.[22]

(d) X's ability to apply P without using rules is dependent in each particular case on some of the properties corresponding to $Q_1 ... Q_n$ being apparent to them.

(e) Whether or not X's are willing to apply P in a particular case, when their state of mind, sense organs, and nervous system are normal, is a function solely of which of the properties corresponding to $Q_1 ... Q_n$ and non-Q_1 ... non-Q_n, are apparent to them.

(f) Each of the properties corresponding to $Q_1 ... Q_n$ is such that either (i) Y's sometimes explicitly recognize and describe it, or (ii) its being apparent to Y's is sometimes necessary for their applying some other predicate without using rules.

VII

Now that we have some idea what to count as differences of conceptual scheme, we can clarify how they are related to differences of language. Particular schemes and particular languages do not go together *necessarily*, because which predicates in a language are grammatically simple and which disjunctive does not determine necessarily which ones speakers can and which they cannot apply without using rules, or which kind of test is given priority. There *could* be a group of people, tribe E, who speak A-language and apply the disjunctive predicates corresponding to our species names without using rules, as we apply our species names, but must use rules for their own species names. They can tell just by looking whether to apply the expression "bird-with-features-a-b-c-...-or-d-e-f-...-or-g-h-i-...-or-..." (thought of as a single run-on unit) to something, but must use a rule to determine whether to call it a "ϕ". Moreover, these may be the tests to which they give priority. If so, E's have the same conceptual scheme that we do, and a different one from A's, despite the fact that E's language is the same as A's.[23] Our species classes are natural and A's are arbitrary for them.

There is, nevertheless, a very close connection between particular languages and particular conceptual schemes. For obvious reasons it is to be expected that the simple predicates in a language will, in general, be the ones which speakers *can* apply without using rules, and hence they are likely to be the ones for which non-rule-applying tests have priority. Complex predicates, we might say, *contain* the rules for applying them; it is clear from the structure of the predicate "bird with features $a, b, c, ...$; or $d, e, f, ...$; or $g, h, i, ...$; or..." what features count in what ways toward its applicability. Simple predicates like "swallow" do not contain inference instructions of this sort; if one uses a rule in applying them he must have learned it independently. It is to be expected that the cases in which a language provides rules of inference will be, by and large, those cases in which rule-applying tests have priority. Tribe E is quite a curiosity. Their language provides them with rules when they are less important and not when they are more important.

VIII

Nothing has been said yet about reality as it *really* is. One might suppose that in order to understand the notion of conceptual schemes, of ways of experiencing things, we must understand how things are *in themselves*, apart from how they are experienced; we must look outside all conceptual schemes to observe the raw "uninterpreted" data, basic properties of things (the "*ding an sich*"), and compare them with facts as "interpreted" by a conceptual scheme, in order to assess the conceptual scheme's contribution. That I do not think this is necessary should be evident from the fact that primary characteristics need not be basic but may be secondary with respect to some other characteristics. The data of people with different schemes need not qualify as "facts in themselves", but may be facts seen or understood in a lower order scheme which both groups share (or in similar aspects of their different schemes). So we can distinguish between the corresponding data, and the different ways in which the groups experience it, without deciding whether those data or anything else are conceptual scheme-independent facts.

This is fortunate, for it would be very difficult to establish that certain facts are of this sort. It is quite possible to operate with concepts of properties thinking they are basic, not even suspecting otherwise, when they are not, as tribes C_1 and C_2 testify. This is not just an academic possibility. We are tempted to call timbres of sounds, e.g., sounding oboeish, "simple" or "indefinable" qualities, i.e., basic characteristics. But they may be construed as secondary with respect to various features of the overtones (which we could discern if we were trained to, though most of us are not). The relative strengths of an oboe-ish sound's overtones are what make it oboe-ish; they are the "cues" that enable us to recognize its oboe-ish-ness. Different oboe-ish sounds – ones produced by different oboes, different reeds, or different players – probably have quite different overtone patterns, so to recognize *just* the overtone patterns would be to recognize only a property which is disjunctive for us common and peculiar to all oboe-ish sounds. Perceiving the timbres of sounds is perceiving various overtone characteristics in a certain way (i.e., "organized into" certain "emergent" simple-for-us properties) even if we do not know it, even if we have no concepts of overtones at all.

But of course there must *be* some properties which are basic, one might think, whether or not we can say which ones they are; the properties we have concepts of, if not basic themselves, must be based ultimately on ones which are – there must be a reality "in itself" even if we cannot find it. But why could not all properties be secondary? A property is secondary if in order to recognize it some one of various combinations of other properties (which we may not have concepts of) must serve as "cues". If this is true of all properties there is either an infinite regress of properties, or classes of properties, each secondary with respect to the succeeding: or there is a circle (or circles) of dependencies, each property being secondary with respect to others, which are secondary with respect to still others, and so on until we arrive at a group of properties including the first one. Neither the regress nor the circularity seems to me to be in the least vicious. Hence I conclude that there need be no basic properties, no properties which do not constitute a way of ordering other properties which is peculiar to a certain conceptual scheme.

IX

I will conclude with several necessarily sketchy suggestions about the possible significance of the notion of conceptual schemes I have been developing. First, I submit that differences of conceptual scheme of the kind discussed can be considerably more drastic, more exotic, of more intrinsic interest, than the simple examples I have used for illustrative purposes. Three points are relevant here. (The first has been anticipated.)

(a) Tribes with different schemes may not realize that properties which are simple for them are based on other data; they may even lack concepts for the data, as do the C tribes. In such cases their respective schemes are so deeply ingrained that not only the data but the functioning of the schemes themselves, is obscured, with the result (as we have seen) that communication and understanding between them would be enormously more difficult than it would be between us and A's, or us and D's, if it is possible at all.

(b) The different properties which are simple for the two tribes may be ones which have a more fundamental place in their lives and thought than bird species and game-ness have in ours. Forces, possibilities, psychological states, and properties of causation, identity (the relational property of being stages of the same table, person, company), and quantity may be involved. All can be construed as based on very complex criteria, but instances of them are rarely if ever identified by appealing to rules. Whorf's obscure claim that the Hopi Indians have a different concept of time from ours may be explainable at least in part in terms of the notion of conceptual schemes we are considering. One piece of evidence Whorf cites is that the Hopi use one set of number words for collections of contemporaneous objects (e.g., three chairs) and a different set for collections of successive objects (e.g., three days).[24] Perhaps the aggregates to which we apply the predicate "three" are seen by the Hopi to share only a disjunctive property, the property named by the disjunction of their two words for three, whereas (I assume) they are seen by us to share a simple property; the class of all groups of three things, contemporary and successive, is natural for us but arbitrary for the Hopi.

(c) There may be far ranging differences in beliefs, reasoning, and behavior which naturally accompany a difference of conceptual scheme of the kind I have described. I presume that people subscribe to, or act and reason on the basis of, some sort of principle of the "uniformity of nature" (e.g., "similar conditions are likely to have similar consequences"). But the application of such a principle depends on what is to be counted as nature proceeding *uniformly*, what things and situations are regarded as *similar*. Hence its application will vary with different conceptual schemes, since people with different schemes recognize different resemblances. People with different schemes can be expected to make different predictions and retrodictions, to act differently accordingly, and to offer different explanations. Put otherwise, I suggest that the predicates people regard as "projectible" in Goodman's sense, as legitimate terms of generalizations inductively supportable by their positive instances, are likely to be the ones they regard as naming simple properties. So which predicates people project, and thus what inductive inferences they make, depends (in part) on their conceptual scheme.[25]

The above will perhaps suffice, abstract and sketchy though it is, to indicate how different conceptual schemes may be at the heart of very great differences in the lives of two groups of people.

A concrete application of the notion of conceptual schemes that looks promising concerns our psychological predicates. In attributing mental states, personality traits, motives, etc. to other people their behavior (verbal and non-verbal) and physical circumstances serve as our data. But we virtually never consult rules in doing so, and non-rule-applying tests clearly have priority. The data may be either unnoticed cues (e.g., subtle details of gestures and facial expressions) or facts we are aware of (e.g., the fact that Jones walked out of a meeting), or both. It is evident also that any of an immense variety of different combinations of behavioral/circumstantial data may lead us to attribute a single psychological property to a person. The behavior and circumstances of angry people in different instances are diverse in the extreme. Thus, being angry is perhaps a simple property for us, which is based on the property, disjunctive for us, of having one among a large (perhaps infinite) number d of combinations of behavioral/circumstantial features (and/or dispositions thereto). This suggests what is wrong, and what right, about behaviorist analyses of psychological properties. Psychological predicates are not just behavioral/circumstantial predicates in disguise. People who recognize only behavior and circumstances would not see a respect in which the many and various cases of anger are alike; instances of anger would constitute an arbitrary class for them. (We might plausibly say that they do not have the concept of anger.) But on the other hand, there is nothing to look for *behind* the behavior and circumstances; seeing what angry persons have in common is just seeing their behavior and circumstances in a certain way. The anger of another person is not some mysterious transcendent property which for some peculiar metaphysical reason we cannot observe "directly", but only via the (questionable) mediation of behavior and circumstances. His behavior and circumstances (including dispositions) might reasonably be said to *constitute* his being angry, through the mediation of our conceptual scheme.

A group of people who recognize the same behavioral/circumstantial data about other people that we do, but not the same psychological properties (different ones, or none at all), would have a different conceptual scheme from ours. Indeed, just such a difference of conceptual scheme obtains between people who admit unconscious mental states and processes and people who do not. For the former, but not for the latter, there is a simple property common and peculiar to all cases of anger, conscious and unconscious.

Whether a theory of psychological states along these lines will ultimately work remains to be established. But it seems a possibility worth pursuing.[26]

The University of Michigan

NOTES

[1] Cf. Whorf (1956).
[2] *Op. cit.*, pp. 240–241, 253.
[3] Strawson (1959) p. 59.
[4] It is conceivable, however, that our color and temperature concepts constitute, in a sense explained below, ways of ordering lower level "data" which the other tribes recognize. If so, their conceptual scheme may be different from ours.

[5] It does not seem likely that anything is to be gained by considering a language which is partly translatable and partly not. The part which is would (it seems) simply constitute different terminology, and the part which is not would, for all we know, just serve to describe facts of kinds we are not aware of.

[6] Some of Whorf's critics seem to have felt this problem. Cf. Black (1962) p. 249; and Cohen (1962) pp. 65–66.

[7] The literature on the Whorf hypothesis abounds with examples of this kind. Cf. Brown (1958) pp. 234–238, 255, 256; Hoijer (1954) p. 96; Lenneberg and Roberts (1956) pp. 25–30; Whorf (1956) pp. 216, 259.

[8] To simplify my example, for reasons that will be evident later, I will assume that bird species are determined solely by features that can be ascertained from a distance at one not too long glance (and/or listening). This leaves out parentage, internal body structure, mating capabilities, etc.

[9] But it need not be. (1) could, e.g., contain infinitely many disjuncts. A question might be raised about how we could learn to use 'swallow' if this is so, for we would have to be able to recognize each of infinitely many combinations of features as being sufficient for making a bird a swallow. But this should be no more mysterious than our ability to apply 'red' correctly in any of an infinite number of different situations.

[10] This conclusion might not be correct for some kinds of properties. If the members of a tribe apply the word 'funny' to a class of things very different from those we apply it to, but in agreement among themselves, and if in addition they generally laugh at what they call 'funny' and not at what they do not, we might well say that they mean by 'funny' what we mean by it. (Who is right about the disputed cases is a difficult question.) The meaning of 'funny' is essentially connected with certain typical ways of reacting toward or treating things. Notice that this cannot be put simply by saying that 'funny' is (partly) "evaluative". If the trible just mentioned value things they call 'funny' roughly to the degree that we value what we call such, but otherwise react to them differently (e.g., suppose they react to them as we do to spine-tingling adventure stories or touching love stories) they still could not be said to mean what we do by 'funny'.

[11] Expressions like 'notice' are *sometimes* used in such a way that we could say one notices (or "unconsciously notices") a bird's features in recognizing it as a swallow, even if he cannot say what they are. But it is also true that we might naturally express the fact that he cannot say what the features are by remarking that he *did not* notice them when he recognized the swallow. I am following the latter usuage.

[12] Indeed it would be impossible to do this if the supposed analyses (e.g., (1)) are infinitely long. Cf. note 9.

[13] For more argument on this matter, see my paper (1963) pp. 357–368.

[14] Quine (1960) suggests a notion of data as simply the physical stimuli (light waves, etc.) impinging on one's body. This notion is considerably weaker than mine, since I require data to be either facts one is aware of or ones which serve as "cues" for others that he is aware of.

[15] I am not considering *con*junctive properties here, such as that of being a red square.

[16] I do not mean to imply that the very same property which is simple for one society might be disjunctive for another. Perhaps the fact that property P is simple for one group and P' is disjunctive for another ensures that P and P' are distinct. Of course the same *class* may be natural for one group and arbitrary for another, if there is a property simple for one group, but none simple for the other, that all and only members of that class possess.

[17] For simplicity, I am ignoring other possible methods of verification, e.g., applying a non-disjunctive rule. I will assume that in all of the cases to be considered verification by applying a disjunctive rule, or by using no rule at all, has clear priority (in the sense sketched below) over any other method of verification. Also, by "rule-applying recognition" I will mean recognition by applying a *disjunctive* rule, and by "non-rule-applying recognition" I will mean recognition without applying a rule of any kind.

[18] This distinction is obviously far from clearcut. Members of a group may differ concerning which test has priority. Or they may simply not have made up their minds; if the two tests rarely or never conflict the question of which has priority may never arise. Also, there may be a rule which is allowed to override non-rule-applying tests in some fringe cases, but which would be rejected entirely if it did not coincide with those tests in most instances. When it is unclear which test has priority it will be unclear whether the property in question is simple for the society. Or, we may say, *to the extent* that non-rule-applying tests have priority the property is simple for the society.

[19] Cf. R. Clark, 'Prima Facie Generalizations', this volume, p. 42.

[20] Identifying games this way goes naturally with what might be called the "verbal-ostensive" manner of teaching concepts. We may teach someone what 'game' means by describing various activities and telling him they are to be called "games" (as well as describing and labelling non-games for him). Eventually he will "grasp the concept" and proceed to use 'game' correctly in other cases. This is very unlike giving a verbal definition; we have not spelled out anything like necessary and sufficient conditions, and need not even have said what conditions are relevant. It is also unlike ordinary ostensive teaching in that there do not have to be actual, or even apparent, instances of the concept to point to.

[21] This guarantees that there will be a class natural for X which is not natural for Y. There could be properties simple for each of two tribes which are based on different data, and hence I suppose are different properties, but which *happen* to be extensionally equivalent. In this case the same classes will be natural for each tribe. One may want to allow this also to count as a difference of conceptual scheme.

[22] This last clause is to ensure that the predicate is not *trivially* disjunctive, as p or p, and (p and q) or p are.

[23] Their language and *A-language* consist of the same marks and sounds with coextensive ranges of application, although perhaps not always with the same *meanings*.

[24] Cf. *op. cit.*, pp. 139–140.

[25] Cf. Goodman (1965). It should be obvious that on my account, blue and green are simple properties and *grue* and *bleen* are disjunctive for English speakers.

[26] I am indebted to Sydney Shoemaker for valuable assistance on my Ph. D. dissertation, from which this paper is extracted, and to John Perry who helped to improve the extraction.

BIBLIOGRAPHY

Black, M., 1962, 'Linguistic Relativity: The Views of Benjamin Lee Whorf', in *Models and Metaphors*, Ithaca, N. Y.

Brown, R., 1958, *Words and Things*, The Free Press, Glencoe, Ill.

Cohen, L. J., 1962, *The Diversity of Meaning*, London.

Goodman, N., 1965, *Fact, Fiction, and Forecast* (2nd ed.), Indianapolis.

Hoijer, H., 1954, 'The Sapir-Whorf Hypothesis', in *Language in Culture* (ed. by H. Hoijer), Chicago.

Lenneberg, E. and Roberts, J., 1956, *International Journal of American Linguistics*, Memoir 13, pp. 25–30.
Quine, W. V. O., 1960, *Word and Object*, Cambridge.
Strawson, P. F., 1959, *Individuals: An Essay in Descriptive Metaphysics*, London.
Walton, K. L., 1963, 'The Dispensability of Perceptual Inferences', *Mind* **82**, 357–68.
Whorf, B., 1956, *Language, Thought and Reality* (ed. by J. B. Carroll), Cambridge.

PAUL ZIFF

SOMETHING ABOUT CONCEPTUAL SCHEMES*

Despite tiresome explications ununderstanding may be rife and rampant. A fulsome detailing of particulars may aggravate without alleviating a failure of communication. Contrariwise, a choice grimace can sometimes suffice to convey even a subtle suggestion. Not prolixity however but an aspect of the etiology of ununderstanding is our immediate concern.

The causes of ununderstanding are legion: any factor relevant to the understanding of what is said may contribute equally to an instance of ununderstanding. Thus phonetic, phonemic, morphologic, syntactic, semantic, discourse and perceptual factors may be operative, may effect a lack or a failure of understanding or even a misunderstanding of what is said. But what I am principally concerned with here are what I shall call "specifications" or, more precisely, "an absence of specifications".

I have in mind certain careless cases: in them communication proceeds succeeds readily easily; no one notices anything out of the ordinary, for nothing is, and no one seems to wonder. A case: a cheetah is a fast feline; such is the common conception, mine too; cheetahs have been clocked at eighty miles an hour; one says then 'A cheetah can outrun a man'. And is one apt to be understood? Of course! But how and why are hard to say or even see.

There are cases of ununderstanding mediated by the use of special words but this is not apt to be one of them. Had one said 'A zibet can outrun a man' some would wonder: 'zibet' is perhaps a rare name for a common civet cat. 'Cheetah' is a known name of a common sort of cat.

Standards criteria and the like sometimes give one pause, occasion genuine ununderstanding. 'Have you been in Kalamazoo?'. I cannot cope with that query without wagging the following tale: 'I was on a train that rolled slowly through Kalamazoo; I looked through the window; I did not get off: have I been in Kalamazoo?'. That is not the way it is with cheetahs outrunning men.

Where words and standards are plain and easy, grammar or syntax or

Pearce and Maynard (eds.), Conceptual Change, 31–41. Reprinted from Paul Ziff: Understanding Understanding. Copyright © 1972 by Cornell University. Used by permission of Cornell University Press.

cooccurrence conundrums may still cause one to come a cropper: 'I couldn't fail to disagree with you less' said the late President Eisenhower tripping a fine linguistic trap. The syntax of 'A cheetah can outrun a man' is as standard as can be.

One says 'A cheetah can outrun a man'. We put it to a test: a man and a cheetah are turned loose in a field. The man lopes away while the cheetah sits lazily in the sun. But 'can' is not 'will'. The cheetah can even if it did not. We try again. This time we force the cheetah to run by beating it; but the man easily outruns the cheetah: the cat encumbered with an awkward two hundred pound weight moves sluggishly.

An encumbered cheetah is a cheetah. And I said a cheetah can outrun a man. Should I have said not that a cheetah can but, more cautiously, that some cheetahs can outrun a man, namely unencumbered cheetahs? But what about an unencumbered cheetah whose feet have been bound since birth? If we make provision for foot-bound unencumbered cheetahs shall we not also have to provide a place for three-legged cheetahs, drugged cheetahs, cheetahs forced to run after being force fed and so on and on? Would it be better then simply to say 'Some – never mind which – cheetahs can outrun a man'?

But communication gropes stumbles and eventually collapses when the forthright 'A cheetah can outrun a man' is so cautiously qualified. It is true that some – never mind which – cheetahs can outrun a man. But it is also true that some – never mind which – men can outrun a cheetah (namely those able ones racing against disabled cheetahs). And it is also true that some – never mind which – cheetahs cannot outrun a man and some – never mind which – men cannot outrun a cheetah. So so far men and cheetahs would seem to be on a par with respect to running. But they are not: a cheetah can outrun a man. Another case: cheetahs don't have horns; what about a cheetah that has been subjected to a successful horn graft? Should we conclude that then anyway some cheetahs don't have horns? Possibly some feel a temptation to switch from 'a cheetah' to 'some cheetahs' when they ponder the remark 'A cheetah can outrun a man'. But not many are likely to feel such a temptation when pondering the remark 'Cheetahs don't have horns'.

These cheetah cases are not curious special or rare. Similar problems are encountered when one considers and ponders such comments as 'A tiger is a large carnivore': isn't a new-born tiger a tiger and yet hardly a large carnivore? 'Skim milk is a healthful food': isn't skim milk laced liberally with strychnine nonetheless skim milk and hardly healthful? 'This car gets thirty five miles to the gallon': what about the leak in the fuel tank?

In each of these cases a difficulty lies lurking in an absence of specifications. One says 'A cheetah' and no further specification of what one is speaking of is given; one says simply 'A tiger', 'Skim milk', 'This car'. And as comments could occasion confusion, so could commands and queries. There is, indeed, no lack of a variety of cases exemplifying an evident absence of specifications, cases in which, though a possible ununderstanding lies lurking, communication usually proceeds smoothly and easily and free of difficulty.

An officer to a private: 'Shut the door!'. 'Sir' says the private and does nothing else. 'Shut the door!' is not the same as 'Shut the door now!'. 'Shut the door now!' said a wily officer whereupon a wilier private responded with 'Sir' at once shutting and immediately reopening the door. 'Shut the door now!' is not the same as 'Shut the door now and leave it shut!'. 'Open the door!' said an officer but a private refused on the grounds that the preceding year the officer's superior had said to him 'Shut the door and leave it shut!' and that order hand not been rescinded.

Another case: students are enjoined not to leave a classroom while a lecturer is there lecturing, and this on pain of expulsion. What if a fire breaks out in the classroom during a lecture: is a student forced to choose between flames and expulsion? In the interests of sanity we could accordingly modify the explicit formulation of the rule to read 'Students are not to leave a classroom while a lecturer is there lecturing unless a fire breaks out'. But if students need not face fire, must they confront an invasion of soldier ants, stay for a sabre-toothed tiger, ignore a sudden infestation of cobras? Or what if a student requires an immediate appendectomy? Or what if a student is heavily bored?

Another case: What time is it? Where? Here of course. When? Now! According to what standard? What time is it here and now according to Eastern Standard time? But when? When the question was first raised or now? And does that mean when I am done or when you last closed your mouth? With what degree of precision? Plus or minus ten hours? And anyway, according to whose timepiece?

One says 'A cheetah can outrun a man' and one says simply 'A cheetah': further specifications of what one is speaking of are usually not given, usually not wanted. Is it simply that further specifications are in fact implicit in the discourse? But cannot what is implicit in discourse be made explicit? Yet when one attempts to detail, to enumerate some set of specifications that would serve to safeguard the truth of a remark like 'A cheetah can outrun a man', the task seems impossible. It is not merely that such set of specifications would have to have indefinitely many and anyway prodigiously many members, but its membership would have to be remarkably heterogeneous. If there were any such set it seems that its vastness and heterogeneity would preclude the possibility of an effective specification of its membership.

A familiar compendious way of coping with cases requiring some specification is by means of an appeal to what is normal or to what is ordinary or common or typical or standard or characteristic and so forth. So one says 'A cat has four legs, a tail, whiskers, at least ordinary ones do'. An appeal to what is ordinary would seem to provide a sufficient specification to safeguard a remark like 'Cheetahs don't have horns'. The query 'What about a cheetah that has undergone a successful horn graft?' can then readily be replied to with 'Ordinary cheetahs don't have horns'.

But what is supposed to count as an "ordinary cheetah"? How many spots does an "ordinary cheetah" have? Can it growl? Can it growl for an hour? a week? a year? Can it whistle "Dixie"? How old is an "ordinary cheetah"? A specification in terms of what is "ordinary" is not apt to be illuminating if what is "ordinary" is itself so singularly difficult to specify.

An appeal to what is ordinary or normal and so forth seems to constitute an adequate safeguarding specification in some cases, partially so in others, and not at all in still others. Thus the comment 'Ordinary cheetahs don't have horns' seems to be true as it stands, is in no need of further specification. 'A tiger is a large carnivore' is only partially firmed up by the specification 'A normal tiger is a large carnivore' for of course adulthood is a further relevant factor. Thus the comment 'A normal adult tiger is a large carnivore' seems to be true as it stands, is in no need of further specification. (But what if there were a machine, a tiger-shrinking device, available such that a normal adult tiger could be shrunk to the size of a newborn tiger? Then I suppose a further specification could be required here. One could say 'A normal unshrunken adult tiger is large carnivore'. Since no such device exists, no such specification is required.)

An appeal to what is ordinary or normal and so forth accomplishes nothing at all in connection with the comment 'A cheetah can outrun a man'. The comment 'A normal cheetah can under normal conditions outrun a normal man' is merely obscure. Cheetahs do not ordinarily typically commonly or even normally race against men or, for that matter against anything. They do chase after and catch antelopes. What are the normal conditions under which a normal cheetah can outrun a normal man? Is the terrain to be rough or smooth? Say it doesn't matter, that either rough or smooth is to count as normal. Then what if the lay of the land is such that it is slow rough broken uneven under the cheetah's feet, flat fast and easy beneath the man's feet? Are these to count as normal conditions? An appeal to normality is bound to be futile when it consists in nothing more than the invocation of a label. Again, normality would appear to be irrelevant in connection with the comment 'This car gets thirty five miles to the gallon'. The vehicle in question might be used for commuting from Long Island to New York and thus typically operated in traffic jams. The relevant specification in this sort of case would require an appeal to optimum operating conditions and to some sort of ideal road conditions under which the vehicle is envisaged as being operated.

In the case of commands injunctions and the like, in place of an otiose appeal to what is normal or typical or common or ordinary, one might hope to formulate some sensible principles with which to parry any thrust arising from an absence of specifications. Thus one might contend that to understand a command is to understand that the command is to be obeyed, that the action that constitutes performance of the command is to be performed, as soon as is feasible. So if one is told to shut a door and there is a door at hand to shut then one is to shut the door at once. Whereas if one is told in the middle of the night to eat three meals a day, one can comply by waiting till breakfast time and beginning then. The view in question would thus be that implicit in explicit injunctions is the enjoinder 'Perform as soon as is feasible!'.

The supposition that there is such an implicit enjoinder is plausible in a few cases, implausible in many. Explicit negative injunctions resist it.

'Don't eat meat!': there is no action that constitutes performance; in consequence one can hardly perform as soon as is feasible. And can one seriously suppose that there is an enjoinder to perform as soon as is feasible implicit in such injunctions as 'Stay awake!', 'Exercise self-restraint!', 'Keep off the grass!'?

Even if one could somehow sensibly suppose that, despite the evident difficulties, implicit to many explicit injunctions is an enjoinder to perform as soon as is feasible, our problems here are scarcely less troublesome than they were: one would still have to invoke still other implicit factors with which to fend off the folly that may be found in the absence of specifications. Thus one would have to conceive of some sort of principle to the effect that one is not to undo what one has done, some principle that would serve to block the move by which, say, after complying with the command to shut a door, one immediately reopened it. And if one succeeded in conceiving and formulating such a principle, still further principles would be needed to cope with still other absurdities that may abound in the absence of specifications. In short, the difficulties inherent in any attempt to furnish some safeguarding set of specifications are equally to be found in any attempt to furnish some safeguarding set of principles.

Appeals to normality and to principles do little to resolve the issues posed by an absence of specifications; appeals to reasonableness, though perhaps self-satisfying, are equally inefficacious. Thus in the case of the rule to the effect that students are not to leave a classroom while a lecturer is there lecturing, instead of attempting to state an unstatable list of exceptions and escape clauses, one could append to the explicit formulation of the rule the clause 'unless there is good and sufficient reason to do so'. And would that make anything clearer? When is there good and sufficient reason to do so and when not? When a fire breaks out, or when there is an invasion of soldier ants, and so on. But the problem still is: how on?

I say 'A cheetah can outrun a man' but what about a cheetah encumbered with an awkward two-hundred-pound weight? It can't outrun a man. Then it is untrue that a cheetah can outrun a man? No. It is true that an encumbered cheetah is a cheetah, and it is also perhaps true that an encumbered cheetah cannot outrun a man. But it does not follow that a cheetah cannot outrun a man. When I say 'A cheetah can outrun a man' am I speaking of an unencumbered cheetah or of an encumbered cheetah? Neither.

If one meets a cheetah on his way then either one meets an unencumbered cheetah or one meets an encumbered cheetah. Either he had a chicken in his mouth or he did not. When it comes to cheetahs in a henhouse, there is no *tertium quid*: the principle of excluded middle rules the roost. But one can speak of a cheetah and think about a cheetah that is neither unencumbered nor encumbered.

A cheetah can outrun man: any cheetah? No. Then only some? No, not that either. A cheetah that can outrun a man is like Hamlet and like Macbeth. Did Hamlet have an aunt? Was Macbeth's left foot larger than his right? These questions go unanswered: there are no answers to give. When I said 'A cheetah can outrun a man' did I mean a cheetah with long white whiskers or one without long white whiskers? Neither one nor the other. But there is no such thing as a real live cheetah that is neither one nor the other. But that only means that if one points at a real live cheetah then one is pointing either at a cheetah with long white whiskers or at a cheetah without long white whiskers. And that is all right here because speaking of a cheetah and pointing at a real live cheetah need not be the same in this respect.

Speaking of a cheetah, as one does when one says 'A cheetah can outrun a man', is like modeling a cheetah in clay or like doing a pictorial representation of a cheetah. Staring at a pictorial respresentation of a man in ordinary black opaque unbulgy riding boots we need not ask: is that man a web-footed or a non-web-footed fellow? A real live man must be one or the other, but a man in a picture is not a real live one.[1]

One has a conception of a cheetah: a long-legged spotted cat, about the size of a small leopard, and having blunt nonretractile claws. And one has a conception of man, and a conception of running, and of outrunning, and of what can or cannot be, and so forth. One's different conceptions of these different matters are related and interrelated in diverse ways; they form what may be called "a conceptual scheme".

I say 'A cheetah can outrun a man': in so doing I indicate that my conception of a cheetah, of running, of outrunning, of a man and so forth all stand in certain specific relations. It is as though a cheetah, a man, running, outrunning and what can be are all different points in some conceptual field: in telling a hearer that a cheetah can outrun a man I offer him, as it were, a bit of a map of a portion of that field indicating certain relations between the indicated points. But if he is to understand what is said then it is up to him to read the map aright. As the woman Crookback says in the *Chuang Tzu*: "It is easier to explain the Way of a sage to someone who has the talent of a sage, you know".

If the hearer is to understand what is said then the hearer, like the speaker, must have some sort of conceptual scheme; he must have some conception of a cheetah, and of a man, of running and so forth. If he conceives of a cheetah as being about the size of a small leopard and as being a long-legged spotted cat and so forth and if his conceptual scheme is of a common kind and his beliefs of a familiar sort and so forth then if thoughts of absent specifications come to plague him in connection with the comment that a cheetah can outrun a man, he will be able to give them the treatment they deserve. For he will see that to say that a cheetah can outrun a man is not to say anything about whiskered or unwhiskered cheetahs. For whether or not a cheetah has whiskers makes no difference at all when what is in question is running ability. Neither is it to say anything significant about encumbered or unencumbered cheetahs. If a cheetah can outrun a man then so can an unencumbered cheetah; whereas if an encumbered cheetah cannot outrun a man that only shows what one knew all along, that an encumbrance is an encumbrance. For he will himself be able to figure out that an animal about the size of a small leopard encumbered with an awkward two-hundred-pound weight would be hard put to outrun a man. And in like vein without exercising excessive ingenuity he should also be able to figure out for himself that the existence of dead cheetahs has no relevant bearing on the truth of the claim that a cheetah can outrun a man.

If a hearer is to understand what is said then not only must the hearer, like the speaker, have some sort of conceptual scheme but he must also understand the form or representation employed by the speaker when the speaker says 'A cheetah can outrun a man'. Cheetahs, ordinary real live cheetahs, have been clocked running at eighty miles an hour. No man has ever run at that speed. Reflecting on these truths and incorporating these statements in one's conceptual scheme may lead one to the related truth that a cheetah can outrun a man. And presumably the speaker in claiming that a cheetah can outrun a man had in mind as exemplars those ordinary real live cheetahs that were clocked running at eighty miles an hour.

But there is of course no necessity here to adopt just that form of representation, to fix on such exemplars. For one's focus could readily be different. Reflecting not on the truths indicated but on other matters, one could state what would seem to be a contradictory view. For suppose it were the case that though we were interested in comparing the running abilities of men and cheetahs, only aged and infirm cheetahs were of interest to us. Assuming that aged and infirm cheetahs tend to be slow and reflecting on these matters and incorporating the appropriate statements in one's conceptual scheme, one could easily be led to incorporate the related truth that a cheetah cannot outrun a man. Another case: a tiger is a large carnivore. What about a newborn tiger? We focus on the adult of the species. We adopt a particular form of representation, a particular form of projection, much as in map making we might, in a particular case to serve our particular needs, prefer a Mercator projection. But we could, were there reason to do so, alter our focus, adopt a different form of projection: a tiger is a small carnivore, for a newborn tiger is small. But then of course what about an adult tiger? No picture captures everything: even the best picture of a cat won't purr.

Focusing and a shifting of focus is readily seen in connection with contrafactuals. For on the assumption that the antecedent of the contrafactual is true, that is, if such-and-such were true, one concludes that the consequent would be true. But the character and truth of the consequent depend on the conceptual scheme in which the antecedent statement is assumed to be incorporated and on the measures taken to accomodate the truth of the antecedent. One says 'If Ceasar were in Viet Nam now, he'd drop the hydrogen bomb'. But shifting focus one could just as well claim 'If Ceasar were in Viet Nam now, he'd use catapults'. I say 'If I were in New York now, then I wouldn't be in Canada'. But I could also say 'If I were in New York now, then I'd be in London and New York at the same time'.

How far can one go here? Could one claim and truly that cheetahs have horns? It is certainly true that cheetahs don't have horns. But is there a shift of focus, a change of one's point of view and a different, a novel, form of representation that would warrant the claim that cheetahs have horns? I don't think that at present there is but I don't see why there couldn't be. For suppose we were cornuphiles and had found a way to give a cheetah horns and suppose we were interested in developing a race of horned cheetahs. These cheetahs yet to be could be our exemplars, could set the standards in terms of which cheetahs are to be characterized. So one perhaps could then say 'A cheetah would be a difficult pet' without supplying or requiring the specification 'horned'; and perhaps one could then say 'People tend to disdain hornless cheetahs'. And one could say 'A cheetah is a horned spotted long-legged cat' and to the query 'What about all the cheetahs to be seen today?' one could reply 'Only inferior hornless cheetahs are to be found at the moment'.

If a hearer is to understand what is said then he must have some sort of conceptual scheme and he must understand and appreciate the form of representation employed by the speaker. How he understands the latter is, I must confess, largely a mystery to me, at least at present. I suspect it will continue to be so for some time. Possibly it has something to do with the fact that a truly understanding hearer is likely to be a speaker as well; possibly for the most part a hearer is apt to assume that a speaker employs the same form of representation as he, the hearer, would were he to be the speaker. Sometimes of course a switch in a form of representation is altogether apparent. Thus *Webster's Seventh New Collegiate Dictionary* writes after 'alpenstock': "a long iron-pointed staff used in mountain climbing". After 'palimpsest' it writes: "writing material (as a parchment or tablet) used two or three times after earlier writing has been erased". One could have a new alpenstock, one that had never been used, but one could not have a palimpsest that had never been used, or at least an unused parchment could not be a palimpsest. Again, after 'doorplate' it is written: "a nameplate on a door". And after 'headache' it is written: "pain in the head". But if a nameplate were removed from the door and put on a shelf it would still be a doorplate. But a headache isn't a headache unless its in the head: even if the pain or ache began in the head and seemed to shift downwards towards one's ankle, one wouldn't have a headache in one's ankle.

When it comes to understanding commands injunctions and so forth it seems plausible to suppose that the hearer takes the speaker to be employing a form of representation such that exemplars of whatever it is that constitutes compliance with the command are to be found in the hearer's experience. Thus if a hearer is told 'Shut the door!' it is as though he were told "Do that which is done such that when it is done you will have obeyed the command "Shut the door!"!'. And if that were so then in a familiar case there would be no question of now or later, of shutting and immediately reopening the door; for such questions do not ordinarily arise, are not likely to arise if one does do that which is ordinarily done.

One speaks and hopes to be understood. Perhaps there is some satisfaction to be found in the fact that both speakers and hearers can contribute their fair share to instances of ununderstanding.[2] At any rate, despite an avid desire to achieve true niggardliness in this respect, I dare say that I have done more than my bit here.

The University of North Carolina
at Chapel Hill

NOTES

[*] This essay appears as Chapter VIII of P. Ziff, *Understanding Understanding*, Ithaca, 1972.
[1] See my *Philosophic Turnings*, Cornell University Press, Ithaca, 1966, Ch. IV.
[2] For an example of ununderstanding attributable to the hearer, see R. Clark, this volume, p. 42.

ROMANE CLARK

PRIMA FACIE GENERALIZATIONS

There are statements and judgements which, *prima facie*, are generalizations but whose truth we maintain in the face of exceptions. 'A cheetah can outrun a man.'[1] Hobbled, injured and drugged cheetahs are cheetahs. Not all of these can outrun any man. Thus, although it is true that a cheetah can outrun a man, it is also true that not all cheetahs can do so. Is this a consistent thing to say?

This is not a consistent thing to say if *prima facie* generalizations are, logically, merely simple generalized material conditionals. 'Anything, if a cheetah, can outrun a man' is inconsistent with the truth that some hobbled cheetah cannot. In what follows I shall in speaking of generalizations mean to refer to *prima facie* generalizations unless another reference is explicitly indicated. By '*prima facie* generalization' we understand nothing more than a statement or judgement which is universal and positive, and whose truth is consistent with the existence of an instance of its subject-class which fails to satisfy its predicate. We have, then, in the occurrence of such generalizations, yet another (familiar) instance in which ordinary thought outruns the resources of classical first-order logic; or, at least, of simple and direct applications of that logic.

Perhaps, however, these generalizations are mendacious, their apparent grammatical forms disguising their real logical natures. It is natural to suppose that these generalizations carry (if only tacitly) certain riders.[2] It is natural to suppose that it is the presence of these riders which make these generalizations compatible with what are now seen to be merely apparent exceptions to them. There are such riders. These fall into several types, the occurrences of instances of which yield distinct kinds of *prima facie* generalizations. It remains open at this point what logical resources are necessary to give explicit formal representation to the various kinds of generalizations which result from the occurrences of such riders.

We might think of what cheetahs *can* do in terms of what *any* cheetah *will* do, the thought suitably qualified. Cheetahs, we might think, will outrun any man when conditions are propitious, or if things are put to a test, or some such. Perhaps the simplest reformulation merely construes 'can' as 'possible... will'; A's can B just in case it is possible that A's will B. But *prima facie* generalizations are safeguarded in other ways as well. They can be construed as carrying modifiers other than the familiar modalities. There are "occurrence-operators": It is usually (generally, frequently, normally, typically,) the case that.... There are "epistemic operators": For all I know it is the case that.... There are modified quantifiers: Approximately (nearly, roughly) all A's.... There are conditional qualifiers: other things equal, (if nothing interferes; if nothing prevents them).

These riders each satisfy the minimal requirement. Generalizations qualified by them are compatible with apparent exceptions. Evidently, were it true that *normally* a cheetah can out run a man, this would be quite compatible with the assertion that some hobbled cheetah cannot. Hobbled cheetahs are not the norm. Evidently, were it true that *nearly all* cheetahs can outrun a man, it might yet be the case that some could not.

It is equally evident, however, that although there are these several species of generalization, instances of which meet the minimal requirement, most of these are quite inappropriate as a paraphrase of the example which initiated our discussion. To say that a cheetah can outrun a man is not, in fact, to say what cheetahs normally, typically usually, generally, or frequently do. It is neither normal nor typical, much less frequent, for cheetahs even to run against men. (Perhaps they never have done so.)

In fact, sifting through the sample above, none of the riders yield a generalization the assertion of which says what probably would be intended by an assertion of our specimen sentence. The epistemic quantifier and simple modal qualifiers are clearly inappropriate. And the conditional qualifiers apparently fail too. 'A cheetah can outrun a man' is not to say what cheetahs will do, other things equal, or unless prevented or interfered with. For, other things equal, an unconstrained cheetah will not outrun a man, but take cover.

Confronted by such common facts, it is tempting to retreat in one of two ways. I believe these retreats to involve false steps, and so I shall mention them only summarily. They suggest, however, an alternative, which is the reason for including them here. The first temptation is to replace our original specimen sentence by another of a certain form. Perhaps we have been too literal. To say that a cheetah can outrun a man is to say something the gist of which comes down to the fact that the cheetah is, after all, by nature faster than man. It is not to say what (some or all) of the existing cheetahs (frequently, usually, or always) actually do. What we need, rather, is to understand the logical character of sentences like 'The Cheetah is faster than Man'.

This suggestion may lead to another, the second temptation. For if the task now is to explicate this replacement sentence, rather than the original, it is natural to ask: What is the point in such assertions? How are categorical generalizations about kinds deployed? And it is natural to answer: To sanction certain inferences and explanations; to validate certain anticipations; to justify certain excuses. We say: "He shouldn't have expected to trap it that way, the Cheetah is faster than Man." Or, "That's a cheetah, so we can't expect to run it down." The categorical generalization about kinds, we may think, is an object-language expression reflecting the presence of a correlated material rule of inference.[3] It is the material rule which ensures the truth of the generalization. (Think of the way in which tautological object-language conditionals match formal rules of inference; Modus Ponens, say.) It is the material rule which explains the role of the generalization in explanations and justifications. It is the rule which explains the immunity of the generalization to apparent counter-instances.

Tempting or not, I believe we should refuse these backward steps. It is wise to resist the second temptation for the fact is that our replacement sentence is *not*, itself, a *rule*, and so it is not a rule of inference. Since it is not, we need in any case to understand its logical nature and to be able to specify its truth-conditions. This is of some importance for, mated as the generalization is to a *material* rule of inference (whatever that may be), it certainly is not itself an analytic statement. The cheetah is by nature, and not by logic, faster than man. The present point then is that, without further fussing with the notion of non-logical rules,[4] we need to note only that their presence does not exempt us from further fussing with their object-language correlates. These correlates are our replacement generalizations. So we are back needing to know something of the logical character and truth-conditions of these generalizations. That, after all, was our original task.

It is wiser, too, to refuse the first step. To say that a cheetah can outrun a man is not simply to say that the Cheetah is by nature faster than Man. The two assertions, although evidently in some way related, are not equivalent. For often what a thing can do is a function of its environment and the world at large as well as of what, or how, it by nature is. What things are, and what, being as they are, they can do are two different things. There remains, then, the original problem. How are we to understand *prima facie* generalizations like that expressed by the use of our original sentence? (A further problem has been added as well. How are these generalizations related to those categorical generalizations which we were tempted to take as their replacements?)

Though we seem to have doubled our problems, there is a gain in this. For we now have a hint how to preserve the original generalization in the face of apparent exceptions. Such generalizations are indeed qualified in a way, say, that the categorical replacement is not. We can see something of the kind of rider required, and of the way it should affect the generalization it rides. What a thing is, given its nature, determines what it can do, in certain circumstances, if nothing prevents it. What it can do, in certain circumstances if nothing prevents it, is what it would do in such a case if it were put to the test. The truth-conditions for 'A cheetah can outrun a man' are presumably the truth conditions for saying that if a cheetah were to race a man, it would, if nothing prevented it, outrun the man. We do not say what it will do, if nothing prevents it, but what it would do. If, contrary to fact, a cheetah were to race a man and nothing prevented it, it would win.

We can see via the explication why it is that apparent exceptions are so easily set to one side. Hobbled, or sick, or baby cheetahs are prevented from outrunning a man, were they to race. They are prevented by constraints, or illness, or physiological immaturity from doing what other cheetahs would, if put to the test, by nature do. We never in our original assertion meant to override such possibilities. We can thus see why the original takes the form it does, and why the tacit rider is what it is. For the original generalization is inherently defeasible. It can be "defeated" in all sorts of ways. There are, compatible with the truth of the generalization, an indefinite number of unspecified ways in which the abilities, tendencies, or capacities, ascribed members of a natural kind can, in the course of circumstance, be thwarted or rendered impotent.

The counterfactual character of the explication makes evident the relation of the ability, tendency, or capacity, of the member of the kind to its nature as a member of that kind. The conditional "prevention clause", the explicit rider of the explication, makes evident the defeasible character of the assertion. It has often been remarked that generalizations like these (and unlike generalized material conditionals) are not revised when "defeated". Rather, in the face of exceptions, they serve to institute new lines of rational inquiry: What were the interfering conditions? (It is difficult to see this response as a rational response to an exception to a generalized material conditional. In this case, the conditional is simply false. The reasonable response would seem to be to revise the conditional, to state more circumspectly its range of application, to bring it into line with reality.)

I do not wish to suggest that the original generalization and its explication do more than coincide in truth-conditions. They are not, I think, in the commonsense meaning of the term, equivalent. In any case, we do not in thinking or saying something like the former, think or say the latter. They do, however, I believe, indeed share the same truth conditions. Seeing what these may be is problematic enough for the explication, but at least clearer than for our original generalization. What, then, are the truth-conditions for the explication?

There is, in specifying the truth-conditions of the explication, a double task. There is the subjunctive component, and there is the special rider as well. In dealing with the subjunctive component, we can borrow from the literature. It was Stalnaker's suggestion to apply modal resources to a familiar way of thinking of conditionals.[5] A conditional is true just when, in a world similar to our own save that the antecedent of the conditional is true of it, the consequent also obtains. This is an attractive line to take with the first of our tasks, the subjunctive component of the explication. It is attractive because it yields a semantical account in line with this familiar intuition about conditionals. And it is attractive in terms of the consequences that emerge. For example, if causal generalizations are viewed in this way, there are certain nice results. Thus, even if striking matches causes them to light, it does not follow that the nonlighting of matches causes them not to be struck. The contrapositive is not a consequence of such a conditional. And the transitivity of the connective similarly fails. If going coatless causes one to catch a cold, and catching colds causes one to be irritable, it nonetheless does not follow that it is going coatless which causes one to be irritable. These are, I think, desirable consequences which nicely match our intuitions.

The question then arises: How similar must an alternative world be to our own for the truth-value of the subjunctive to be invariant with the truth of the statement matching its consequent clause in that alternative world? It was Stalnaker's further suggestion[5] to undercut some of these difficulties by requiring that, of the alternative worlds in which the proposition matching the antecedent of the subjunctive is true, we select that alternative which is maximally like our own. It is the truth-value of the consequent in this selected world which determines the truth-value of the conditional. It is not, of course, obvious just what world will, in a given case, be maximally like our own. It is not even clear, without further principles for ordering such alternatives, that there is such a maximal. The main difficulty, however, is not that such principles of order may seem ad hoc, merely invoked as required in different cases to make the theory go. The complaint lies rather with the appeal to a single, maximally similar, alternative no matter how it is picked out. Consider, for instance, arbitrary true propositions, P and Q. Their conjunction implies that if P were the case, Q would be. for the world maximally like our own, in which it is the case that P, clearly is our own since P, by hypothesis, is true. But (in that world), Q by assumption is also true. Therefore the subjunctive is, on this theory, true as well. For arbitrary, unrelated, P and Q, this is an anomalous result. Note, too, that for any proposition, P, and any proposition Q, on the present theory either it is the case that if P were true, Q would be, or else it is the case that if P were true, not-Q would be. But since P and Q may bear no relevance one to the other, many of us would be inclined to deny both of these alternatives. These are, I think, undesirable consequences, which fail to match our intuitions. Our task, then, is to preserve the desirable consequences of this approach to conditionals, while avoiding undesirable consequences like these last.

These last consequences, forcing relevance between unrelated sentences as they do, can be skirted. It is not necessary, in determining the truth-conditions for a conditional, that we select the world maximally like our own in which the antecedent obtains. It is enough instead that we consider only those which are sufficiently so. The conditional, 'If P were the case, Q would be', is true provided Q is true of all those P-worlds sufficiently like our own. 'If P were the case, Q would be' is false provided Q is false of some world sufficiently like our own in which P is true. The question arises: When is a world sufficiently like our own? It is not, I suppose, possible to answer this in general for every proposition and independent of any context. But certain minimal conditions must be met. For instance, we shall certainly wish to hold that, relative to any proposition which actually obtains, our actual world is one of the worlds sufficiently like our own. And we shall wish to isolate contradictory propositions. Since there is no (consistently describable) world in which the contradictory proposition obtains, there is no set of worlds sufficiently like our own relative to these. (The theory, thus, is a theory of a form of conditional and not of the relation of entailment.)

Some structure can be given these informal considerations in the following way. Commonsense conditionals (e.g., of the form 'Were P, would Q') may be symbolized as 'P/Q'. 'P' is the *numerator* of the conditional, 'Q' its *denominator*.[6] We think of these formalizations as mundane sentences, true or false of the actual world. To each such mundane sentence there exists a certain, unique transcription which no longer contains the slash and which is simply an expression of standard first-order logic supplemented with some specific predicates and constants and supplemented with some specific assumptions governing these predicates. Mundane logical truths have transcriptions which are theorems in the resulting extended, first-order logic.

'Twp' says that the proposition P is true in w. '$Rpww^*$' says that, relative to the proposition P, w^* is an alternative of w; i.e., 'w^* is possible relative to w with respect to P;' or, that w^* is, relative to P, sufficiently like w. The transcription of the mundane 'P/Q' is developed in two steps. First, the assertion is relativized to the world in which it obtains, in this case the actual world, o. We have '$To(P/Q)$'. The truth-in-a-world predicate is then confined in accord with the following stipulation:

$$To(P/Q) \quad \text{to} \quad (w)(Rpow \to Twq).$$

More generally, we assume:

$$Tw(P/Q) \quad \text{to} \quad (w')(Rpww' \to Tw'q),$$
$$Tw-(P/Q) \quad \text{to} \quad (Ew')(Rpww' \& -Tw'q),$$
$$Tw(P \# Q) \quad \text{to} \quad (Twp) \# (Twq),$$

for any binary truth functional connective, #, and, for negation,

$$Tw - P \text{ to } -Twp.$$

Depending upon the assumptions made concerning the relation R, various theorems emerge. If we think of R as expressing the basic relation obtaining between worlds sufficiently similar relative to a proposition, it is natural to assume, e.g., that if a proposition is true of a world, then that world is one of those sufficiently like itself relative to that proposition. I.e., we assume that

$$(w)(Twp \to Rpww).$$

On this simple assumption, it turns out (as we should hope) that the following mundane sentence has a transcription which is a theorem of our extended, first-order quantification theory:

$$(P/Q) \to (P \to Q).$$

This is a theorem of the logic of conditionals since

$$(w)(Twp \to Rpww) \to ((w)(Rpow \to Twq) \to (Top \to Toq))$$

is a theorem of standard, first-order logic.

Analogues of the Kripke relations among modal alternatives turn up as well, not surprisingly. For example, just as the mundane modal assertion linking necessity with truth,

$$\Box P \to P,$$

holds in a Kripke-semantics in which the alternativeness relation among possible worlds is reflexive, so too the conditional (now unhappily) links with its consequent if we assume that $(w)(Rpww)$. Assuming this,

$$P/Q \to Q$$

is then a theorem of the conditional logic. Evidently, then, it is implausible to suppose that for any (consistent) proposition every world is possible with respect to itself relative to that arbitrary proposition. (The proposition may, after all, not be true.)

Just as necessity implies possibility, on the assumption that, for every world, there exists an alternative, so too we have

$$P/Q \to -(P/-Q)$$

provided that, relative to a consistent proposition, for every world there exists some world sufficiently like it. I.e., the mundane sentence is a theorem of conditional logics if we assume that

$$(w)\, (Ew')\, (Rpww').$$

A kind of transitivity result follows, without further assumption. We have,

$$P/Q \to P/(P/Q).$$

And on the (plausible) assumption that $(w)\, (Rpow \to R(p/q)ow)$, we have

$$(P/Q)/Q \to P/Q.$$

Conversely, on the assumption that $(w)\, (R(p/q)ow \to Rpow)$, we have

$$P/Q \to (P/Q)/Q.$$

Certain numerator-denominator relationships hold. e.g.,

$$P/(Q\ \&\ R) \to (P/Q)\ \&\ (P/R),$$
$$(P \vee Q)/R \to (P/R)\ \&\ (Q/R),$$

are theorems, while, for instance,

$$P/(Q \vee R) \to (P/Q) \vee (P/R)$$

is not. This last failure, it seems to me, is a highly desirable failure. (It is, however, derivable in Stalnaker's system.)

So a fabric of relationships emerges which has a certain plausibility on an understanding of the conditional like this. The present point, however, is that certain features crucial to our understanding of the conditional are retained on this version, (as they are on Stalnaker's), while certain other unprepossessing features drop by the way. It turns out, as we should like it to, that neither the transitivity of the conditional connective, nor the contrapositive of the conditional, obtain. We do not have

$$P/Q \to (Q/R \to P/R),$$

as a theorem, nor do we have

$$P/Q \to -Q/-P$$

as a theorem.

Equally important, certain further anomalous results are not forthcoming. It does not follow from the fact that the cat is on the mat and the North wind is bitter that if the cat were on the mat, the North wind would be bitter. It is not, i.e., a theorem that

$$(P \ \& \ Q) \rightarrow (P/Q).$$

Nor is it a theorem that

$$(P/Q) \vee (P/-Q).$$

This is nice for, tautologies to one side, we wish our conditional to retain something of the relevance which obtains between the clauses of true, commonsense, subjunctive conditionals.

Our original problem, we recall, was to analyze *prima facie* generalizations. These were generalizations whose truth might be unaffected by apparent counter-instantiations. For one species of such generalizations, (certain generalizations of the form 'A's can B'), we thought of construing them in terms of what A's would do unless prevented. The hope was that truth-conditions for the latter might yield some insight into truth-conditions for the original generalizations, and so, perhaps, even yield some further understanding of their meaning. Considering what A's would do unless prevented lead us to a sketch of the subjunctive element which was present in the paraphrase of the original *prima facie* generalization. There remains next the task of characterizing, and placing, the qualifier which tacitly rides the generalization. "If cheetahs were to race men, they would unless prevented outrun them." What is the effect of the rider, 'unless prevented'? (There is, of course, a range of conditional qualifiers, of which 'unless prevented' is merely one instance. Finer distinctions than are appropriate to our present discussion can be drawn among these. What a thing would do if nothing prevented it need not coincide with what a thing would do were other things equal or if nothing interfered with it. We ignore these differences here.)

'Unless prevented' is an exclusion clause. It exempts the generalization from defeat by overriding circumstances, circumstances the presence of which would ensure an exception to the general truth. To say that if a cheetah were to race a man, then, were nothing to interfere, it would outrun him, is perhaps to say something of this form: $P/(R/Q)$, where 'R' is a subjunctive exclusion clause. If so it is to say something doubly subjunctive; an iteration of subjunctive slashes. It is worth stressing here that it is a matter of some formal importance to our task to determine whether, and where, in our formalization the iteration correctly occurs. It matters whether the iteration occurs, for it is not true (despite clouded commonsense intuitions and the example of the material conditional) that it is logically indifferent whether we formalize our paraphrase of the original generalization as $(P \& R)/Q$ instead of $P/(R/Q)$. The latter is a weaker proposition than the former. It is entailed by, but does not entail, the former on a certain sort of transitivity assumption.

$$(P \& R)/Q \rightarrow P/(R/Q)$$

is a theorem in a system for conditionals which makes an assumption concerning sufficient similarity across worlds like this:

$$(w)(w')(w'')(Rpww' \& Rqw'w'' \rightarrow R(p \& q)ww'').$$

But the converse,

$$P/(R/Q) \rightarrow (P \& R)/Q,$$

is not a theorem and no plausible assumption makes it so.

Equally, it matters where the iteration, if iteration there be, occurs. $P/(R/Q)$ is implied by, but does not imply, $(P/R)/Q$ under a certain transitivity assumption:

$$(w)(w')(w'')(Rpww' \& Rqw'w'' \rightarrow R(p/q)ww'').$$

It is not difficult to choose between $P/(R/Q)$ and $(P/R)/Q$ as possible ways to formalize the paraphrase of our original generalization. The latter says that a cheetah would outrun a man, were it the case that, were they to race, nothing would interfere. But the racing itself guarantees neither that nothing will interfere (nor that anything will for that matter), and so the latter clearly will not do.

It is more difficult to choose between $(P \& R)/Q$ and $P/(R/Q)$ as formalizations. Shall we choose the stronger, or the weaker, proposition (if we choose either)? There are natural inclinations to try the weakest of assumptions, other things equal. And there are, it seems to me, slight intuitive or commonsensical considerations which push in this direction as well. But these will seem at best moot to anyone who is already committed the other way. The fact is that so far the rider which qualifies the subjunctive has been hidden in its formalization by a single letter. But what does the rider actually come to? Will not that determine, finally, where we place the occurrence of the rider in the formalization of the paraphrase? Probably it should. But there is an embarrassment in the attempt to hazard remarks on the form of the rider. It is not obvious that our present resources are at all adequate to that task. We have at present merely the resources of standard logic supplemented by a sketch of a theory of conditionals. 'A's would B, if nothing were to interfere' suggests that A's would B if there were to exist no circumstances which, in contunction with A, would ensure that not-B. This seems in the right direction, but there is in the suggestion a richness which outruns our present powers of expression. For one thing, 'ensures' needs making out. And there seems need to quantify over circumstances as well. No doubt if these matters were adequately sorted out we could indeed decide where the rider is to go, and why. But lacking these, as we do, our intuitions remain unsupported either by compelling reasons or an adequate formal explication.

Let us suppose, however, as a final speculation, that the subjunctive itself can provide some reasonable approximation to the required sense of 'ensures'. And let us pull down the apparatus of quantification with propositional variables, naively, and without further technical elaboration. We can, then, give an approximate expression for our rider. 'A's would B, if nothing were to interfere' is to say that A's would B, if $-(EC)$ $(A \& C/-B)$; i.e., if (C) $(A \& C/B)$, for consistent $A \& C$. This approximation finally provides some grounds for choosing the position of the rider in the paraphrase. The alternatives now read this way: $(A \& (C)$ $(A \& C/B))/B$, on the one hand, and $A/((C) (A \& C/B)/B)$, on the other. The former is true just in case B is true of any world sufficiently like our own in which two truths obtain: A is the case and were A true with any other proposition, B would be as well. That is to say, with respect to our original example, that cheetahs would outrun men were they to race them and were their racing in any circumstances at all enough to ensure that they would outrun them. But this is too strong; evidently worlds like that, in which the generalization cannot be defeated are not sufficiently like our own.

The second alternative requires instead that in A-worlds sufficiently like our own a certain subjunctive must hold. It must be true in them that, were it the case that A together with any circumstance would ensure that B, then B would be the case. That is to say, with respect to the original example, that were cheetahs to race men, then they would outrun them were it the case that they would outrun them in all circumstances in which they were to race. I.e., that none of those were circumstances in which the conditions conspired to ensure they would not. By casting the rider into the denominator of the main clause of the subjunctive, the ideals in which nothing interferes are separated from worlds like our own in which things may. We have then, in our speculation, presumptive evidence (not for the form of the rider itself, but) for the position in which the rider should appear. The upshot is this: *Prima facie* generalizations of a certain form (which carry if only tacitly certain conditional qualifiers) have the same truth-conditions as have certain iterated, subjunctive conditionals. The sentence 'A's can B' has the same truth-conditions as has the generalized subjunctive: If x were to A, then, were nothing to interfere, it would B. And this, finally, has the form: $A/(R/B)$, the truth-conditions for which are articulated in a logic of conditionals.

Indiana University

NOTES

[1] Ziff, P., 'Something About Conceptual Schemes', this volume, p. 31. The discussion makes clear that what is meant is 'Cheetahs can outrun men' or 'The Cheetah can outrun Man'.

[2] See Achinstein (1965).

[3] See, e.g., Sellars (1953).

[4] The notion of a material rule, and of its relation to a mated generalization, has never (to my knowledge) been made clear. An earlier expression of skepticism on all this is recorded in Clark (1956).

[5] Stalnaker (1968).

[6] We wish to distinguish the clauses of subjunctive conditionals from sentences. The latter, e.g., but not the former, occur perhaps as the antecedents or consequents of material conditionals.

This is an idiom I borrow from Robert Binkley. More important, many of the points in the sketch which follows grow from discussions and suggestions of his which go back to a summer in 1967 and an academic year together at the University of Western Ontario, 1968–69. I doubt, however, that he would wish to be associated with what is said here or shares the motivation for saying it.

BIBLIOGRAPHY

Achinstein, P., 1965, '"Defeasible" Problems', *The Journal of Philosophy* **62**, 629–33.
Clark, R., 1956, 'Natural Inference', *Mind* **65**, 455–72.
Sellars, W., 1953, 'Inference and Meaning', *Mind* **62**, 313–38.
Stalnaker, R. C., 1968, 'A Theory of Conditionals', *Studies in Logical Theory* (ed. by N. Rescher). *American Philosophical Quarterly:* Monograph Series, Monograph No. 2 Basil Blackwell, Oxford, pp. 98–112.

ROBERT W. BINKLEY

CHANGE OF BELIEF OR CHANGE OF MEANING?

I should like to begin by noting that while I regard it as a considerable privilege that I should be addressing you at all this evening, it is to me a still greater one that I should be addressing you first rather than second. This is an order of speakers quite the reverse of what the printed program might have led you to suspect, and I want to comment on that to forestall a possible misapprehension. It is well known that the philosophy department here at the University of Western Ontario, and in particular its chairman, Robert Butts, and its Colloquium organiser, Glenn Pearce, are free creative spirits in no way restrained by the iron hand of custom. It might be thought therefore, that this reversal of speakers results from an undisciplined urge to novelty for its own sake. "Let us be the first", you might be thinking they said to each other, "to have the commentary before rather than after the main paper". I want to assure you that it is no such thing. We are all aware that the main speaker of the evening, Professor Sellars, has published many papers dealing in general with the topic of our colloquium, *Conceptual Change*, and in particular with the topic of this evening's session, *change of belief or change of meaning*? All of that is to be taken as the main paper of the evening, and assumed read. My task is to comment upon it; Professor Sellars' will then be to reply to my remarks, though the possibility cannot be ruled out that before the evening is over yet another card will have slipped from his sleeve onto the table.

Now a second introductory remark. I do not take up this task in the spirit of a hostile critic. Professor Sellars' philosophical territory is the one in which I grew up, and which I find most familiar, and far from wanting to do battle over it, I would prefer to take visitors on guided tours. That will, in fact, be the main purpose of my remarks; to set before you the views of Professor Sellars on our topic insofar as they can be extracted from his writings, simplifying and streamlining them, however, in such a way as to bring to the center of our attention certain problems and puzzles that remain to be resolved.

I shall not, however, undertake to do this in the most direct way, which would be by way of a careful exegesis of selected texts. Instead, I shall present an independent development of the theme, using my own terminology for the most part rather than that of Sellars, but reaching a position that I believe to be essentially equivalent to his.

And so to work. Our topic is "Change of belief or change of meaning?" This is not just a disjunction; it is a question. Some change has occurred, and we wish to know what kind of change it was. We think that, for some reason, we can narrow it down to these two possibilities; it was either a change of belief or a change of meaning, but we do not know which. (Of course, it might have been both, but we try not to think about that.) Some such background as this seems to be presupposed by our title, and I shall work from it. For there is an argument of some persuasiveness to show that this is bound to be a silly and idle question, one which, if it can be answered at all, can be answered easily.

Notice, to begin with, that the changes we are considering are changes in different things. It is *words*, or more generally, *linguistic expressions*, that undergo change of meaning. It is persons who undergo change of belief. And this by itself should be enough to arouse suspicions about our question. For it now becomes, in part, the question, "What has changed, a word or a person?", and it is hard to see how one could be perplexed about a thing like that. It will be necessary, however, to consider each of these changes in more detail.

First, words. What are they, and how can they change in meaning? The basic notion, of course, is that of a word having a meaning at a time, since change of meaning occurs when the same word has different meanings at different times. (For convenience, I shall here treat having no meaning as a limited case of meaning.) Thus we must consider the form "Word W has meaning M at time t".

It might be suggested at this point that I should add "in language L for person P". I shall resist both these suggestions, at least for the time being. I do not wish to relativize to the language because I shall regard words in different languages as different words, however much alike they may look. And I do not wish to relativize to persons because I think our concern should be with persons who use words in their standard public meanings. But I will return to these points later, and will then relent a little.

The words themselves, I think we must say, are universals; their instances are events, events of language users using language. Such events ordinarily consist in the production of a sound or inscription of of a certain design by the language user but there are other possibilities. A word instance in a gesture language, for example, might consist in assuming a certain posture, or in moving in a certain way.[1]

It is easy to say *roughly* what constitutes the identity of a word, that is, to say what the rule is for telling whether or not an event is an instance of a particular word. Roughly, it is a matter of the shape or other physical properties of the inscription or sound produced. But it is not entirely that. The production of the shape must be the using of the language, and this is not simply a matter of the shape produced. The maker of donuts is not making '*O*'s. Further, identical shapes may be associated with different words, and dissimilar shapes with the same word. 'Felt', the matted fabric, and 'felt', the past participle of 'feel', are different words. Chaucer's 'bygynneth' and our 'begins' are the same word. For this and other reasons it is not easy to say *exactly* what constitutes the identity of a word, and it is fortunate that I will not have to on the present occasion.

It should be clear from the above that I am thinking of a language that can itself undergo changes, and have a history without losing its identity. I can thus allow myself to speak of language L as it is at time t, and as it is at time t', rather than speaking of two distinct languages. And so a word can change meaning without thereby becoming another word, or its language another language.

Words, then, are universals. What sense can it make to speak of a universal possessing a property, such as that of meaning a certain thing, *at a time*? A universal has a property at a time if something can be said about all possible instances of the universal at that time. The way to say something about all possible instances is to use the subjunctive. Accordingly, to say that a universal has a property at a time is to say that any instance of the universal occurring at that time would have a certain related property. And in our case, to say that a word has a meaning at a time, in a sense of 'meaning' appropriate to word-universals, is to say that any instance of the word occurring at that time would have a certain meaning, in a sense of 'meaning' appropriate to word-instances.

We come down then to the question what it is for a word-instance to have a meaning, and there I shall leave the matter for the moment. I shall simply remark that having a certain meaning is an *essential* property of a word-instance in the sense that *it* could not have *its* meaning unless it were true that *any* instance of the word would have the same meaning.

Let us turn instead to the other change we must consider, change of belief. The root notion is that of a person having a belief at a time. Change of belief occurs when a person has a certain belief at one time but not at another. Beliefs are to be identified by their contents, that is, the thing believed; same content, same belief. So we need to know what it is for a person to have a belief of a certain content at a time.

On this point, Sellars holds that having the belief is having a proneness at that time to mental acts of a kind appropriate to the belief content. The belief is a potency; its actualization is a mental act. If there is behavior associated with a belief, it is behavior that we think flows from these mental acts. These belief actualizations are to be conceived by analogy with the use of language; in particular, with the assertions of sentences in spontaneous unpremeditated discourse; sentences, that is, of a suitably indicative as opposed to imperative or interrogative character.[2]

The content of a belief is given by a sentence the use of which constitutes the basis for the analogy. I call such a sentence a *proper vehicle* of the belief. Normally, there will be more than one proper vehicle for a belief; one, in fact, for every language in which the content of the belief can be expressed. And even a single language may have more than one proper vehicle for a given belief if it provides several inter-translatable ways of expressing the belief.

Our concern, however, is not with proper vehicles in general, but rather with the proper vehicle in *our* language, for it is through that sentence that we understand what the content of the belief is. This, of course, will not be the proper vehicle that the believer would use to express his belief unless he happens to share our language. And even if he does, it might not be the sentence that he would select. Some people, for example, are inarticulate, and cannot summon up the resources of their language to express their beliefs in any satisfactory way. Indeed, it is not even necessary that the believer *have* a language containing a proper vehicle for the belief, whether he can formulate it or not. Some people are creative and have beliefs beyond the power of their language to express. Perhaps there are believers who possess no language at all

– the pet cats and dogs, for example, of anti-linguistic philosophers.[3] But none of that is of any interest to us, except in so far as we are concerned with the question how to find out what the beliefs of others are. What matters is that *we* should possess a proper vehicle: given that, we can say what the believer's belief is, whether he is able to or not.

For we construct descriptions of beliefs out of the proper vehicles for them. Thus Jones may have a belief of such a kind that we decide that "The earth is round" is our proper vehicle for it. We *could just describe* the belief by, "Jones has a belief whose proper vehicle in our language is 'The earth is round'". Ordinary language, however, would prefer us to say "Jones believes that the earth is round". Subject to a few qualifications to be mentioned below, I am prepared to regard these two descriptions logically equivalent. That is, I am prepared to say:

> Jones believes that $p =_{Df}$ Jones has a belief whose proper vehicle in our language is "p".

Or better, bringing in another dimension of the account,

> Jones believes that $p =_{Df}$ Jones is prone to mental acts analogous to the assertion in our language of "p".

Sellars, of course, does not put the point in just this way. He would prefer, using his dot-quotation convention, to form a verb out of each sentence of our language, a verb describing the act of using that sentence, or any equivalent sentence in any other language, and then to use that verb to say, for example, "Jones is prone to· The earth is round·ing in his mind", where the "in his mind" indicates that the use of the verb is analogical. But I think these two ways of speaking come to much the same thing.[4]

Now for the qualifications. The first set have to do with the fact that often for one reason or another we cannot come by a proper vehicle in our language, and must settle for an approximation, or for some less than completely definite specification of a proper vehicle. For example, Jones' belief may not be exactly that the earth is round; he may have a somewhat more determinate belief than this because his language contains a word, not matched by any word in our language, covering a range of non-cornered shapes narrower than that covered by our word 'round' but broader than any of our more determinate words such as 'spherical', or 'ellipsoidal'. Knowing this, we may still offer "Jones believes that the earth is round" as a description of the belief because we allow ourselves to build belief descriptions not only upon proper vehicles, but also upon our best approximations to proper vehicles.

This is, I think, the most common case. We ordinarily describe a belief not strictly by offering a proper vehicle, but rather by indicating in various ways, and with more or less precision, what a proper vehicle in our language would be like. There are two indefinitenesses involved here, and they should be kept distinct, even though ordinary language does not provide an easy means of doing so. On the one hand, Jones is indefinite to some extent about the shape of the earth; on the other, we are indefinite to some extent about Jones' thought. What we want, of course, is to be definite about Jones' indefiniteness. Failing that, we approximate. The ordinary sentence, "Jones believes that the earth has some kind of round shape", is inadequate since it does not make clear whether it is us or Jones who is saying 'some kind of' instead of saying which. A more satisfactory formulation is

> $(\exists \emptyset)$ ('$\emptyset$' is a more determinate predicate under round and a more determinable predicate over 'spherical' and Jones has a belief whose proper vehicle in our language is 'The earth is $\emptyset$').

In this formulation, the quantification is substitutional, and the variable ranges in the substitutional way over the predicates of our language. Not, however, simply over the predicates that are in our language *now*, but rather over those together with any other predicates that our language will or might come to have in the future.[5] Our language, that is, might come to contain an exact proper vehicle for Jones' belief, and our formulation gives us some notion as to what such a predicate would be like. This quantification into quotation marks might strike some logicians as unseemly. I have argued elsewhere, however, that it is all right to quantify substitutionally into quotation marks provided that the quantifier does not simultaneously bind an occurrence of the variable outside the quotation, a condition which is met by the present formulation.[5]

The qualification just considered might be said to concern the *sense* of beliefs; there are also difficulties concerning their *reference*. I shall treat these under two heads, problems concerning the *object* of reference, and problems concerning the *mode* of reference.

First, the object. The proper vehicles in our language may involve referring expressions, that is, expressions whose job in language, at least in part, is that of securing a connection between the sentences in which they appear and some object in the world. One of the questions that arise concerning sentences containing such expressions, and so concerning the beliefs for which they are the proper vehicles, is whether the expression really does secure the connection as it is supposed to do. The key point to be noticed here is that one of the presuppositions *I* make concerning *my* use of *my* language is that *my* referring expressions really do succeed. This presupposition is, in a relevant sense, part of the language. It would be wrong of me to make serious use of a referring expression if I thought it did not succeed, wrong as violating a rule governing the use of language in discussion which I believe Sellars would call a *dialectical* rule.[6] But while referential success is dialectically presupposed by my use of my referring expressions, I do not make the same pressupposition about the use of referring expressions by others. This gives rise to difficulties.

For one thing, the other fellow's referring expression may not, in my view, succeed in establishing a connection with any object at all. In that case, I will feel uneasy in describing his belief by the use of a proper vehicle containing that referring expression, for to do so would seem to implicate me in the use of that expression, and so suggest that *I* am presupposing its referential success. Thus, even though "The Loch Ness monster swims deep" may be a proper vehicle for a belief of Jones', I will hesitate to say that Jones believes that the Loch Ness monster swims deep for fear of being accused of thinking that there is such a thing as the Loch Ness monster.

This hesitation can be overcome, though at the cost of some violence done to ordinary language, by clinging resolutely to the principle that just as the phrase "X believes that" placed in front of a sentence cancels any suggestion that the sentence is being asserted as true, so too it cancels any dialectical presupposing about the success of the referring expressions it contains.

A related point arises when we consider *which* object is referred to by a referring expression, assuming it does refer to something. When we become convinced that two of our referring expressions connect their sentences to the same object, we express this conviction by placing the identity symbol (or the 'is' of identity) between them. Since the two expressions connect their sentences to the same object, it will make no difference which is used when we are using sentences to say what properties the connected object has. In these cases, therefore, we allow the one referring expression to be replaced by the other, according to what has come to be called Leibniz's Law. When, however, we are using a sentence as a proper vehicle in the construction of a description of someone's belief, any concern with the connected object has been cancelled by the 'X believes that' out in front, and Leibniz's Law does not apply.

The fundamental fact here is that, despite appearances, sentences of the form "X believes that Y is F" do not assert a relation to obtain between X and Y; in fact, Y is not referred to at all. This fact becomes more intelligible when we remind ourselves of the essentially meta-level character of our use of proper vehicles in belief descriptions. From the fact that "Y is F" is a proper vehicle it does not follow that the expression 'Y' refers. Nor does it follow that a sentence formed from "Y is F" by replacing the 'Y' with another expression co-referential with it will also be a proper vehicle.

But, it will be asked, if "Jones believes that Smith is rich" does not *assert* a relation between Jones and Smith, may there not still *be* some relation between them when Jones does have this belief? There are, I think, two relevant relations that may be present, though neither of them *need* be present. These relations correspond to two different senses in which referring expressions may be said to *connect* their sentences with their objects.

In one sense, we may call the connection a semantical connection. (Sellars, I think, would prefer to say *pseudo-connection*.) It is this sense that is involved when someone asks me to what object a referring expression connects its objects, and I tell him. I tell him by using a referring expression of mine that, as I assume, is known to him and is itself connected to the right object. That is, I simply put into correspondence the questioned referring expression and another that is taken to be unquestioned. If he asks me to what object 'the author of *Waverley*' connects its sentences, I tell him by saying, in effect, "'The author of *Waverley*' is co-referential with *our* referring expression 'Scott'". This is the semantical connection, and it is not really a connection between the referring expression and the object, even though ordinary language would have us use the form "'The author of *Waverley*' refers to Scott".

Still, we can use this semantical connection to affirm a relation of sorts between Jones and the man Smith he believes to be rich. (Sellars would prefer to say a *pseudo-relation*.) For we can say:

> "Smith is rich" is a proper vehicle for a belief of Jones', and 'Smith' as used here is co-referential with *our* 'Smith'.

By the use of 'our' before the last 'Smith', I restore the referential assumptions that are cancelled when I merely describe belief. A more interesting case is the existential generalization from this:

> ($\exists e$) ("e is rich" is a proper vehicle of a belief of Jones' and 'e' is co-referential with our 'Smith'),

where 'e' ranges in the substitutional way over actual and possible referring expressions of our language. This gives the transparent sense of belief that "Smith is believed by Jones to be rich" is supposed to capture.[7]

The second kind of relation between a believer and some of the "objects" of his belief I shall call the *real* relation. This relation arises from the fact that language users and their uses of language are in the same world as the objects referred to. Let us consider a language user who is describing his world in some not totally inadequate way, and let us consider the stream of speech flowing from him not as speech but simply as physical phenomenon. We would expect to find, I think, two relationships holding between that phenomenon and the surrounding world. First, we would expect there to be an isomorphism of some kind so that bits of the speech could be paired with bits of the world. Second, we would expect there to be a causal relationship permitting us to say that the speech is the way it is *because*, in part, the world is the way *it* is; had the world been different, the speech would have been different. In the absence of these relationships, it is hard to think what grounds we could have for saying that the language user is describing *his* world.

A key feature in both the isomorphism and the causal relationship will be the use of at least some of the referring expressions. These referring expressions will be paired by the isomorphism with particular *objects* in the world. (Other features of the speech will be paired not with objects, but with how the objects are, or are related.) And these referring expressions will be causally related to the objects with which they are paired in the sense that the causal explanation of the occurrence of the referring expression in the speech will be, in part, the fact that the object acted causally on Jones. This causal influence may be remote and tenuous, and may be mediated through other language users. Socrates, for example, exerts a causal influence upon us of the relevant kind because his direct effect on Plato and others has led to our having his name in our vocabularies.

This relation of causality and isomorphism connecting some referring expressions with their objects provides the basis for a relation between a believer and the object. For when Jones believes that Smith is rich, it may be that Jones' referring expression for Smith, or its mental analogue, is related in this way to Smith. If so, then in having this belief Jones is related to Smith in a real way, definable in terms of isomorphism and causality; related in a way in which he would not be if he did not have this belief. This relation is an important one for epistemology and other things: the point to be stressed here however is that it is not the relation that is being discussed when we say that Jones believes that Smith is rich, nor even when we say, in a transparent way, that Smith is believed by Jones to be rich.[8]

I shall now move on from the problems associated with the object of reference to those associated with the mode of reference. Under this head I have in mind the difficulties that arise from the fact that some of our referring expressions employ token-reflexive devices. These are referring expressions whose mode of reference is such that an expression instance will connect its sentence to one or another object depending on the circumstances in which it occurs. Such an expression, while admirably suited for use by Jones in his circumstances for expressing one of his beliefs, may very well not be suitable for use by us in our circumstances for describing Jones as having that belief. Jones, for example, might come to have a belief which he would then express by saying, "I am ruined". What are we to use as a proper vehicle here upon which to build our description of Jones' belief? "I am ruined" does not seem right since it suggests reference to me rather than to Jones; certainly, "Jones believes that I am ruined" will not do. "He is ruined" will not do either, in spite of the fact that ordinary language suggests it by offering the form "Jones believes that he is ruined". The whole pathos of the situation lies in the fact that Jones is thinking of himself in the first person in this belief, a feature that this formulation does not capture.

The form "Jones believes that he himself is ruined" is better insofar as it suggests this, but ambiguities arise in more complex cases such as "Smith believes that Jones believes that he himself is ruined", ambiguities that have been fully mapped by Castañeda.[9] "Jones is ruined", of course, will not do for a proper vehicle since Jones may not be aware that that is his name.

The solution, I think, is to abandon ordinary language and its clumsy attempts to convert token-reflexives appropriate to the believer into token-reflexives appropriate to the belief describer. The proper vehicle is "I am ruined"; we describe the belief by saying that these are the words in our language appropriate for use *by Jones* to express his belief. But if we cling to this line, as I think we should, then we will sometimes have to do more in describing a belief than cite the proper vehicle. If the proper vehicle involves token-reflexives whose reference depends on circumstances, then we will have to provide information about the circumstances. There is no problem here for the 'I' since the relevant information is provided when we say whose belief it is we are describing. Other cases are more complex. Suppose that Jones observes Smith, a perfect stranger, walking down the street carrying a sign with a political slogan, and Jones acquires a belief that he could then express by saying "That man is a fool". Our policy requires us to say that our proper vehicle for that belief is just those words. But to describe the belief fully we must indicate the circumstances in which these would be the right words to use for expressing the belief. Something like the following, perhaps, will do:

> Jones has a belief whose proper vehicle in our language is "That man is a fool", 'that man' being taken in demonstrative reference to a man observed by Jones on such-and-such an occasion.

(I should perhaps pause to observe that I am here moving a bit beyond anything that I find in Professor Sellars' writings.)

A special case of this of particular importance to us is that of temporal reference and the use of tenses. Such phrases as 'now', 'two days ago' as well as the tenses themselves are token-reflexive devices, and so, by the strategy just adopted, their use in the proper vehicles by which beliefs are described will need to be accompanied by a commentary characterizing the circumstances. Thus if on Monday Jones looks out the window and acquires a belief which he would then express by saying "It is raining", then we at a later time would describe that belief by saying,

> On Monday Jones had a belief whose proper vehicle is "It is raining".

Here we do not need a further indication of the temporal reference since the present tense in the proper vehicle always points to the believer's *now*, that is, the time of the belief, just as the first person always points to the believer.

On Tuesday, the situation becomes more complicated. Jones, let us suppose, retains his belief of the previous day about the weather of the previous day. He can (on Tuesday) express that belief by saying "It rained yesterday", and so we set out to describe his belief by saying, On Tuesday Jones had a belief whose proper vehicle is "It rained yesterday".

So far, so good; Tuesday's 'yesterday' must point to Monday. But this formulation overlooks the fact that what we have on Tuesday is simply the second day in the life of Jones' belief. For I think we want to say that it *is* the same belief that Jones has on the two days; Jones has not in any way changed his mind about Monday's weather. We can include this aspect if, when giving the proper vehicle, we specify the time at which it is the proper vehicle. If we do this, then we can say,

> On both days Jones had a belief whose Monday proper vehicle is "It is raining" and whose Tuesday proper vehicle is "It rained yesterday".

The assumption here is that the same belief can continue over time, clothing itself in different proper vehicles at different times, and that the change of clothing is not change of belief. However, in order to be said to retain the same belief, Jones must keep track of his temporal location. If on Tuesday he remembers that it rained but not how many days ago, then his belief has changed. The Tuesday proper vehicle is not "It rained yesterday", but simply "It has rained"; and the Monday proper vehicle for *that* belief would not be "It is raining" but rather "It is raining, has rained or will rain".[10]

We may now return to our main topic, "Change of belief or change of meaning?". We have seen, up to a point, what it is for a word to have a meaning at a time and what it is for a person to have a belief at a time, and so what the corresponding changes are, and we are in a position to consider the objection raised at the outset that the question "Change of belief or change of meaning?" must be a silly one. The argument would be something like this: The only words we are considering are those constituting the proper vehicles for beliefs, and the only proper vehicles we are considering are those used by us on the single occasion on which we try to describe someone's belief, or change of belief. When I say that Jones used to believe X but now believes Y, the X and Y are expressed in *my* words with the meanings these words have for *me now*. So there is no room for change of meaning. It may be, of course, that Jones' language has been undergoing rapid change; it may even be that the words he used to use to express his belief that X are now the words that he uses to express his belief that Y. If so, then there has been a simultaneous change of belief and meaning for Jones. But this presents no *problem. If* we know what he believes and used to believe and what his words mean and used to mean, then we have no difficulty in characterizing the situation. And if we do not, then we have no hope of getting clear about it until we do. The important questions are "What do these words mean?", "What does this person believe?", not "What has changed, the meaning or the belief?".

The answer to this objection lies in noticing that the connection between meaning and belief is much closer than the objection supposes. To understand this connection, we must inquire further into what it is for a word to have a meaning at a time. We left this point above with the claim that for a word to have a certain meaning at a time is for it to be true that *any* instance at the time of the word-universal *would have* that meaning. To move beyond this we must reflect on the use of the subjunctive here, and its implication of lawfulness. What sort of laws?

Here is the point, I think, to introduce Professor Sellars' well known view that the meaning of a word in a language is like the role of a piece in a game; for example, like the role of the pawn in chess. A language is like a game in that it is governed by rules, and these rules determine the roles of the pieces and the meanings of the words.

The rules of the language state what would be correct or incorrect uses of language in circumstances of various kinds. These are primarily rules of criticism rather than rules of action. They do not seek to guide our use of language directly, for we do not ordinarily use language in conscious obedience to a rule; we just say what comes to mind. This is particularly true of the spontaneous use of language which is the foundation of the analogy through which we conceive the mental actions which are the actualizations of beliefs. The rules rather guide my use of language indirectly by directly guiding the activities of my language trainers (of whom I may be one) in their efforts to instill proper language habits in me.

These rules are the laws lying behind our use of the subjunctive when we say that any instance of the word-universal would have a certain meaning. To be an instance of the word universal, an event would have to be a manifestation of the rules, and so have the relevant meaning.

Sellars has provided a rough classification of the various kinds of rules of language. In addition to the formation rules, there are rules governing three different kinds of transition. First, there are language-entry transitions, and so rules specifying the correct linguistic response to developments in the environment. Second, there are language-departure transitions, and so rules specifying the correct behavioral response to linguistic developments. Some of these language-entry and language-departure transitions involve going up and down between object and meta-language, and so give the misleading appearance of belonging to the third category, language-language transitions. This is the case, for example, with transitions between "'p' is true" and "p", as well some of the stages of inductive reasoning to be considered below. The language-language rules proper govern linguistic responses to linguistic developments, all at the same level of language. These last, the transformation rules, may be further subdivided into three classes. First, there are logical rules of deductive inference. Second, there are consequence rules which govern the reformulations required by the fact that the language user moves about in space and time, and so must speak from different points of view. The tense transformations referred to above in connection with the life of a belief over time are an example of the operation of consequence rules. Finally, and most important for our purposes, there are the so-called P-rules of physical entailment. These are rules of non-logical inference, and are mirrored in the object language by law-like statements.[11]

A complete set of such rules determines a language game or conceptual system, and fixes the meaning of every term in the language. Changing these rules changes the meanings. In principle, I suppose, changing any rule implies changing all meanings, since everything interconnects. But one can think of at least two reasons why, for practical purposes, the meaning change resulting from a change in the rules can be limited to the relatively small circle of concepts directly involved. First, there is a limit to our powers to discriminate differences in meaning. This stems from the fact that when we try to say what the meaning of a term is we do so by relating the term to an equivalent term presumed understood, and since these equivalences may be only approximate, a single understood term may serve as rough equivalent for several distinct terms, actual or possible, the differences thus being unremarked. There is, of course, in principle the possibility of specifying all meanings exactly by listing all the rules, but that is beyond our power. And anyhow, there is the second reason, which is that in fact our language is not completely determined by rules. This is the open texture phenomenon. Because of this, a change in one part of the language might be damped out before it ramified through all the rest. (Here again I am perhaps moving a bit beyond anything that Sellars has actually said.)

Meanings change, then, when the rules defining our conceptual system change. And according to Sellars, this is a frequent occurrence. It occurs, for example, whenever a new law-like statement comes to be accepted, or a previously accepted one abandoned. It occurs whenever a proposition is shifted from the category of the contingent to that of the necessary, or vice versa. And, in a more profound way, it occurs whenever the conceptual framework of common sense gives way to that of theoretical science.

In connection with this last point we must note that on Professor Sellars' view, our present conceptual system is actually a complex one composed of two main subsystems which co-exist in our minds under conditions of cold war tension. These he has named the manifest and scientific images of man in the world. The manifest image is the conceptual system of common sense as purified and extended by philosophical analysis and inductive science. The scientific image is the system emerging from theoretical science which postulates a domain of unobserved entities to which the world of the manifest image stands as appearance to reality. These two substructures are at present linked by yet another kind of rule, the correspondence rules. But this is a temporary methodological expedient since the scientific framework, which has developed out of the manifest one, is destined ultimately and completely to replace it. In the mean time we have both. We are like a second generation pioneer family living in the parental log cabin while we build a grand new modern house for ourselves beside it. Some of us, perhaps Paul Feyerabend, are eager to move into the new house at once. Others, perhaps Ernest Nagel, maintain that people cannot really live in anything but log cabins, and that the new house can only serve as a workshop from which we will always have to return to the cabin to eat and sleep. Professor Sellars holds that we can and should move into the new house eventually, but that it would be wiser to wait until it is finished. He is particularly anxious to wait until the painters are through, since he does not want to live in a house where there are no colors.[12]

For Sellars, the giving way of the common sense framework in favor of the scientific one lies in the future. For this reason, the changes of meaning involved in that shift, while of the most profound kind, are not our immediate concern. More important for us are the changes of meaning involved in the development of the two frameworks, and their subframeworks, considered, up to a point, in isolation. Here the changes to consider are changes of the P-rules. Within this frame of reference, we are now in a position to see more clearly why the question, change of belief or change of meaning?, presents a problem. For the change of meaning we are considering can also be described as change of belief, if for no other reason than that different sentences are asserted as the result of it. This is true in particular of the law-like sentences which Sellars would describe as themselves mirroring the P-rules in the object language; that is, as being statements of the P-rules in the material mode of speech.

For there are two kinds of change of belief. A first, paradigm, kind does not involve change of meaning. For example, Jones believes for a while that Alberta adjoins Manitoba. Then he consults a map and a change of belief occurs. He now believes that Saskatchewan intervenes between them. Changes of this kind can be said to be paradigm because we have a whole language game of its own devoted to them. This is the language game of epistemic appraisal. It is the game in which we express agreement or disagreement, argue with people, estimate the rationality of belief, etc. Logic, conceived as the canon of rational belief, is the basic law of this game. And the whole point of it is making sure that the rules we have got are obeyed, not changing the rules. It acts as referee, not as rules committee, and it is thus designed for use in a context of fixed meanings.

But, as we have just noticed, there are also changes of belief that do involve change of meaning. A chief reason why these changes seem problematic to us in a way in which changes of the first kind do not is that in these cases we also use our system of epistemic appraisal in spite of the fact that it is designed for fixed meanings because it is the only system of epistemic appraisal we have got. For example, suppose that Jones believes for a time that smoking does not cause cancer, and then finally on the basis of such evidence as he has, concludes that it does. We easily use the language of epistemic appraisal here. Jones has acquired a new belief; now he disagrees with the leaders of the tobacco industry who have a different and contradictory belief. We ask whether his new belief is reasonable given his evidence, and so on. Yet, on Sellars' account, while this is a change of belief, it is also a change of meaning. "Smoking causes cancer" is a law-like statement, and is to be treated as a principle of inference, a P-rule. What Jones has done is to change the rules of the language, at least his language, and where rules change meanings change, if Sellars is right. This is a relaxing of our decision to consider only speakers using words in their standard public meaning, but only a very small one. For Jones does not entertain a belief of this sort as sheer idiosyncrasy; he would rather hold that any rational being acquainted with the evidence ought to agree with him. We might put this by saying that in Jones one apprehends a linguistic insurrection whose object is the total overthrow of established language on this point.

A change of belief of this kind, involving as it does a change of meaning, presents two main problems. First, how is the change to be described; second, how is it to be justified. And both these problems are really aspects of a single larger problem; how is our system of epistemic appraisal to be applied in contexts of changing meanings.

Sellars has said a number of things that have a bearing on this. In 'Counterfactuals, Dispositions and the Causal Modalities' he says two things. First, he admits that the change of meaning involved is not one of "explicit definition", even though the terms (in our example, 'smoking' and 'cancer') now "involve" each other in a way that they did not before (p. 288). The use of the terms is now "enriched" (p. 287) in such a way that knowledge of the connection between smoking and cancer is a necessary condition of understanding these terms in their new use (p. 287). Of course, and this is Sellars' second point, Jones used the old unenriched meanings in assembling and assessing the inductive evidence for his new belief (p. 287). Because of this, the old meanings cannot be entirely lost sight of. Induction involves risk of error, and Jones needs to bear in mind that he may be mistaken about smoking and cancer. The way to bear this in mind is for him to keep the old meanings alive in some way so that if the new meanings prove to be ill-advised, the old meanings will be there to retreat to. This "line of retreat", and from the other point of view, a "plan of advance", is part of the logic of the terms (p. 288).

We may notice two points here. First, the shifting of P-rules does not affect the explicit definitions of the terms, which implies that there is a hard core of meaning, given by explicit definition, that is constant with respect to this change. This hard core would be cracked, I suppose, if Jones should come to think that smoking does not after all involve the drawing in of smoke.

Such a hard core is one way of meeting a condition that must be met by any meaning change conforming to the pattern now under discussion. This is the condition that there must be enough in common between the before and after phases for us to say that it is the same word throughout. If this condition is not met, then we are not changing meanings, but replacing one vocabulary with another. It is the difference between changing the rules of chess and giving up chess for football. And the distinction remains even though it is possible to conceive of a series of tiny changes that would gradually transform chess into football.

Second, and more important, we may notice that Sellars is suggesting that we do not work with just a single stage of the developing conceptual system at a time. We work rather with several generations at once; while one will be the current favorite, we nevertheless keep an eye at all times on its ancestors and possible descendents.

But these observations, though helpful in their way, only sharpen our difficulty. For we are reminded on the one hand of the connection between meaning and understanding, and on the other of the connection between law-like statements and induction. Putting these two connections together leads to the following dilemma, concerning the justification of the change. Law-like statements are to be justified by induction. The use of induction presumably means that Jones should have available an inductive argument whose premises will say that such and such a number of people smoked and got cancer, etc., etc., and whose conclusion will say that smoking causes cancer. The trouble is that the old meanings of the terms are used in the premises and the new in the conclusion, and this sounds like the fallacy of equivocation.

That the new meanings appear in the conclusion is precisely Sellars' point. That the old meanings must appear in the premises follows from the fact that an inductive argument is supposed to justify its conclusion, which implies that its premises must be capable of being accepted *and so understood* by people who do not yet accept the conclusion. How else could the argument win these people over?

Sellars' reply to this, I think, would be that it rests on an overly crude conception of inductive reasoning, a conception engendered, doubtless, by the attempt to employ the referee epistemic appraisal system without noticing that meanings are being changed. An inductive argument, for Sellars, is not like a deductive one that sweeps through from premises to conclusion all on the same level. It is rather a complex piece of reasoning that moves on both the object-language and meta-language levels, and employs both the modes of theoretical and of practical reason. The crux of it is that when Jones seeks to justify his law-like statement to Smith, and to win Smith over, he does so by trying to convince Smith that they both have reason to want to change the use of language in such a way as to provide a connection between 'smoking' and 'cancer'. It is thus at bottom practical reasoning culminating in a decision to change meanings. Jones does this by showing that this change is required if their desire to possess a conceptual system that will do as good a job as possible of explaining and predicting what goes on in the world is to be satisfied. The idea of equivocation has no application to an argument with a complex structure such as this. The crucial terms are not *used* at all in the conclusion; rather, they are *mentioned* in an injunction to change their meanings.[13]

The justification, then, of change of belief which involves change of meaning can be understood by appreciating the complex structure of inductive reasoning. But there remains the difficulty of describing this change. This is the problem of how at the same time we can have in our language proper vehicles for both the before and after stages. How, that is to say, can we play two different language games at once?

With respect to this problem, the essential move is made by Sellars in *Science and Metaphysics*. Translated into my terms, it can be put as follows. We saw very early on that we cannot always find an exact proper vehicle in our language upon which to build a description of belief, and so must often settle for more of less indefinite specifications of what such a vehicle would be like. That is what must be done here. But it can be done systematically, since meanings that have been lost can always be regained. Thus, suppose that we now agree with Jones that smoking causes cancer. Then we have no problem in describing Jones' present belief; we simply say that Jones has a belief whose proper vehicle in our language is "Smoking causes cancer". The problem lies in describing the old belief. But we can describe the old belief by taking advantage of the fact that we do not have to *give* a proper vehicle for the described belief. To actually have such a vehicle, we would have to reintroduce into our current language terms having the earlier meanings of 'smoking' and 'cancer'. But instead of doing that, we can say "Some sentence in a possible development of our language translates the old 'smoking does not cause cancer', and is a proper vehicle for a belief that Jones had." (The possible development, of course, is that of restoring the old meanings.) That is,

$$(\exists s) \text{ ('}s\text{' translates the old "Smoking does not cause cancer" and '}s\text{' is a proper vehicle for a past belief of Jones')}$$

this being substitutional quantification of the kind considered above.

The problem, then, in the question, Change of meaning or change of belief? lies in the fact that in some interesting cases the answer is, Both. And the solution lies in the fact that, with a little ingenuity and analysis, our methods of describing and appraising beliefs can be extended to these cases.

University of Western Ontario

NOTES

[1] The notion of a word as an *event* universal is to be found, for example, in Sellars (1969a).
[2] It is hardly necessary to give references to prove that this is more or less what Sellars thinks beliefs are. These points, and the ones to come, can all be found in Sellars (1968) but they have been made over the years in numerous places.
[3] These last statements are an embellishment on Sellars' text. I think they are consistent with what he says, though I do not think that he has actually said them.
[4] On these points one may consult Sellars (1969a) cited above.
[5] Binkley (1970).
[6] See for example, Sellars (1968) p. 123, as well as the earlier paper, Sellars (1953).
[7] On these points one may consult Sellars (1969) in addition to his numerous "'rot' means red" discussions.
[8] In these last few paragraphs I have been discussing as aspect of what Sellars calls the picturing relation between language and the world. It is discussed in Sellars (1968) and also in Sellars (1962a) and Sellars (1962d), both of which are reprinted in Sellars (1963b). The basic idea that semantical concepts operate wholly at the meta-level, and that the connection between language and the world lies in the interaction between the world and the language user, appears even in the earliest papers, written in those distant days when the way of words could still be described as new. For example, see Sellars (1948b).
[9] Castañeda (1967).
[10] Sellars has not, I think, said exactly this. But something of the sort seems to be implied by his discussion of a super-inscriber who keeps track of his position in space and time by means of heartbeats and steps in Sellars (1962d) cited above. See also Sellars (1962c).
[11] These ideas are all to be found in Sellars (1968), but are to be found also in many earlier papers. Sellars (1954) and reprinted in Sellars (1963b) is particularly important. On *P*-rules one may consult Sellars (1958). See also the very early paper, Sellars (1948a).
[12] These points are made in Sellars (1968), and also in Sellars' papers on the philosophy of science, especially Sellars (1962b) and reprinted in Sellars (1963b). See also Sellars (1965). The point about colors is also discussed in Sellars (1956) and in Sellars (1967) both reprinted in Sellars (1963b).
[13] The main reference for these points is Sellars (1964).

BIBLIOGRAPHY

Binkley, R., 1970, 'Quantifying, Quotation and a Paradox', *Nous* **4**, 271–277.
Castañeda, H.-N., 1967, 'Indicators and Quasi-indicators', *American Philosophical Quarterly* **4**, 35–100.
Sellars, W., 1948a, 'Concepts as Involving Laws and Inconceivable without Them', *Philosophy of Science* **15**, 287–315.
Sellars, W., 1948b, 'Realism and the New Way of Words', *Philosophy and Phenomenological Research* **8**, 601–34.
Sellars, W., 1953, 'Presupposing', *Philosophical Review* **63**, 197–213.
Sellars, W., 1954, 'Some Reflections on Language Games', *Philosophy of Science* **21**, 204–28.
Sellars, W., 1956, 'Empiricism and the Philosophy of Mind', in *Minnesota Studies in*

the Philosophy of Science, Vol. I (ed. by H. Feigl and M. Scriven), Minneapolis, pp. 253–329.

Sellars, W., 1958, 'Counterfactuals, Dispositions and the Causal Modalities', in *Minnesota Studies in the Philosophy of Science*, Vol. II (ed. by H. Feigl, M. Scriven, and well) Minneapolis, pp. 225–308.

Sellars, W., 1962a, 'Naming and Saying', *Philosophy of Science* **29**, 7–26.

Sellars, W., 1962b, 'Philosophy and the Scientific Image of Man', in *Frontiers of Science and Philosophy* (ed. by R. Colodny), Vol. I, Univ. of Pittsburgh Press, Pittsburgh, pp. 35–78.

Sellars, W., 1962c, 'Time and the World Order' in *Minnesota Studies in the Philosophy of Science*, Vol. III (ed. by H. Feigl and G. Maxwell), Univ. of Minnesota Press, Minneapolis, pp. 527–616.

Sellars, W., 1962d, 'Truth and "Correspondence"', *Journal of Philosophy* **59**, 29–54.

Sellars, W., 1963a, 'Abstract Entities', *Review of Metaphysics* **16**, 627–71.

Sellars, W., 1963b, *Science, Perception and Reality*, Routledge & Kegan Paul, London.

Sellars, W., 1964, 'Induction as Vindication', *Philosophy of Science* **31**, 197–231.

Sellars, W., 1965, 'Scientific Realism or Irenic Instrumentalism' in *Proceedings of the Boston Colloquium on the Philosophy of Science*, Vol. II (ed. by R. S. Cohen and M. W. Wartofsky), New York, pp. 171–204.

Sellars, W., 1967, 'Phenomenalism', in *Intentionality, Minds and Perception* (ed. by H.-N. Casteñeda), Wayne State Univ. Press, Detroit, pp. 215–74.

Sellars, W., 1968, *Science and Metaphysics*, Routledge & Kegan Paul, London.

Sellars, W., 1969a, 'Metaphysics and the Concept of a Person', in *The Logical Way of Doing Things* (ed. by Karel Lambert), Yale University Press, New Haven and London, pp. 219–52.

Sellars, W., 1969b, 'Some Problems about Belief', in *Philosophical Logic* (ed. by J. W. Davis, D. J. Hockney and W. K. Wilson), D. Reidel, Dordrecht-Holland, pp. 46–65.

WILFRID SELLARS

CONCEPTUAL CHANGE

I

1. When philosophers compare the "before" and "after" of a scientific revolution, thus the birth of special relativity, they tend to divide into two camps: those who speak of a revolution in belief and those who speak of a conceptual revolution. The disagreement is a typically philosophical one, since what is at issue is not the facts but rather the categories in terms of which their deep connections with other facts are to be made manifest.

2. The former or "change of belief" theorists obviously do not deny that an expression which stands at one time for a certain concept may come to stand for another closely related concept. But they do not find in this fact a key to understanding what takes place in the evolution or revolution of scientific theories. As for their own positive views, they may put them in a number of ways, each of which carries a different dialectical burden.

3. They may begin by claiming that relativity physicists have simply come to have different beliefs about length, mass, velocity, etc. Thus physicists have not changed their concept of length, rather they have come to believe, as they did not before, that length is a function of relative velocity. This way of putting it, however, raises the sensitive question as to what it is to have a belief about *length* (*simultaneity*, *velocity*, etc.) – abstract entities all.

4. Those who are inclined towards instrumentalism may put their point by stressing that scientists have come to have different beliefs about the outcomes of certain experiments and operations of measurement. To the extent that instrumentalism is based on the idea that perceptual qualities and relations are simply given, and that the corresponding predicates are their names or labels, it construes conceptual change as the building of

new definitional structures from conceptual elements which themselves do not change.

5. Theoretical terms, on the other hand, for an instrumentalism thus grounded, would "stand for concepts" in quite a different sense. Such "meaning" as they have would be their "use" as elements in a calculus. In the case of some of these terms, this calculational use would involve their occurrence in "correspondence rules". Here they would be cheek by jowl with predicates which stand for concepts proper. But although this cohabitation gives their use an empirical dimension, it does not place them in the definitional hierachy which begins with *labeled, given attributes and relations*.

6. The instrumentalist will cheerfully acknowledge that the "meaning" of theoretical terms changes with each change in the postulates of the theory. For this "meaning" simply *is* the calculational use and, leaving aside the subject matter independent syntax of purely logical and mathematical manipulation, the *distinctive* use of a theoretical term is constituted by the extra-logical postulates in which it appears or, for defined terms, in which their *definientia* appear. Notice that the instrumentalist need not boggle at the concept of likeness of "meaning", nor at the idea that as a theory develops some of its terms have their "meaning" relatively unchanged, whereas others undergo a radical shift in "meaning", the instrumentalist, however, is not likely to be impressed with such scholastic distinctions, for, after all, the "meaning" in question is not *real* meaning. A predicate has *real* meaning when it *stands for* a kind or attribute or relation. And these are entities some of which, e.g. the shape of this table, are directly experienced.

7. Thus the "change of belief" theorist who takes the instrumentalist tack allows all kinds of "conceptual change" as theories develop, but limits the role of conceptual change *proper* to the relatively uninteresting fact that a word which stands for one complex of properties at one time can come to stand for another, perhaps closely related, complex at another. And, as noted before, he does not find the idea of such an occurrence illuminating with respect to the structure of scientific revolutions.

8. Now change of belief theorists need not be instrumentalists, and, since I believe instrumentalism, construed along the above lines, to be false, I shall turn my attention to those who regard themselves as taking a realistic stance toward "scientific objects".

9. It is time, however, to take an initial look at the opposing "camp". Its adherents find it illuminating to say: (a) The expressions 'l', 'm', 'v', 'simultaneous', etc., of relativity kinematics stand for different concepts than do the corresponding expressions of classical kinematics. (b) The new concepts are not simply *other than* the old ones, for each *resembles* its predecessors in some respects and *differs* in others. Thus the new concept expressed by 'l' is, like the old one, a length concept; the new concept expressed by 'simultaneous' is, like the old, a simultaneity concept. (c) Hence we can say not just that these expressions have acquired new meanings, but that they have undergone a meaning change.

10. The exact nature of the difference between "acquiring a meaning" and "undergoing a change of meaning" remains to be spelled out, of course, but it is obvious that the operative word is 'change' – as contrasted with 'replacement' – and the clue to its importance, the conceptual *similarity* implied by such phrases as 'classical length concept' and 'relativity length concept'.

II

11. I emphasized above that "change of belief" theorists need not be instrumentalists with respect to theories. It is sometimes thought that the once standard but now controversial "reconstruction" of theoretical explanation along the lines of Reichenbach, Carnap, Hempel and Nagel, with its three-fold distinction between two conceptual frameworks, one observational,[1] one theoretical, and a set of bridges between them, leads inevitably, once its implications are worked out, however irenically, to the semantical and ontological dualism which is the substance of instrumentalism. I have argued elsewhere[2] that this is not the case, and shall not reargue the matter here. I shall therefore draw upon this familiar account in order to describe in greater detail the issues at stake between our two schools of thought.

12. I shall assume, then, that we can reconstruct, for philosophical purposes, the clash between special relativity and classical kinematics in terms of a competition to see which framework can generate, with appropriate correspondence rules, a more adequate set of experimental generalizations in a common observation framework. Consider two propositions, one from each framework, which contain one or more terms in common from the following list: 'l', 'm', 'c', 'v', 't', etc. In a preanalytic sense these shared terms stand for the same quantities in the two theories. Thus, the working scientist would be comfortable about saying that 'l' in both propositions stands for length, 'v' for velocity, etc.

13. Of course, two competing theories *need* not overlap so neatly in their theoretical vocabulary. Sometimes a theory is replaced by a theory so different in its structure that one would have little temptation to say that any term in the one stands for the same quantity as any term in the other.[3]

14. According to our regimented picture, when two theories compete with respect to the same domain of phenomena, different predictions are made about phenomena at the observation level. Here one can speak with relative ease[4] about a difference in belief involving the same conceptual materials. The confrontation is between those who wish to say that scientists, without changing their concepts of length, mass, etc. *at the theoretical level*, have different *beliefs* about how the length, mass, etc., of objects varies with relative velocity than do their Newtonian counterparts, on the one hand, and those who find it illuminating to say that the difference in belief *constitutes* a difference in the concepts of length, mass, etc., involved.

15. Notice that in the above formulation I attributed to the "change of belief" theorist the view that scientists have changed their beliefs about the length, mass, etc., of objects. This was, however, a bit unfair, for what he wants to claim is that two theoretical propositions, one classical, one relativistic, involving the variables 'l' and 'v', express different beliefs *not* about such abstract entities as *length* or *velocity*, but rather about how the correct answers to the questions 'how long is such and such an object?' or 'how fast does it move?' vary with respect to how fast the inertial frame to which the object belongs is moving with respect to the inertial frame in which the measurements are made.

16. In any event, what he wishes to claim is that in whatever sense predicates express concepts, '*l*', '*v*', etc., stand for the same concepts in relativity kinematics and in Newtonian mechanics. We would have a case of conflicting beliefs involving the same concepts. The conflict would be, in the tough sense, a logical one, and no 'mere' matter of competition for the same job.⁵

17. What about the 'change of concept' theorist? He certainly does not deny that different and conflicting beliefs about the length of objects are involved in the two theories. He argues, however, that '*l*' stands for different, but similar, concepts in the two theories, and that the similarity is a matter of its standing, in each case, for a different *length concept* or concept *of length*.

18. It will not have passed unnoticed that I have been assuming that the "change of belief" theorist is willing to speak of predicates as *standing for concepts* and to use such phrases as 'the same concept' and 'different concepts'. This, of course, many philosophers who stress change of belief are unwilling to do – not because they regard talk about concepts as meaningless, for they would certainly admit that it is just as meaningful to talk about concepts as to talk about beliefs. In effect they challenge the "change of concept" theorist to make useful sense of such phrases as 'stand for different, but similar, concepts' and to explain how one distinguishes between cases in which two conflicting beliefs involve the same concept, and cases in which the conflicting beliefs involve different but similar concepts. To satisfy their demands, one must (a) give criteria for conceptual identity; (b) explicate what it is for concepts to be, in the relevant sense, "similar".

III

19. The word 'concept' has a number of uses, some of which carry a heavy burden of philosophical theory. In one of its less controversial uses, a concept is something a person has, namely a certain ability, and is always the concept *of* something, where 'something' is used in the broadest possible sense. To illustrate, Jones has the concept of triangularity if and only if Jones is able to think thoughts to the effect that such and such an item is triangular. Again, Jones has the concept of conjunction if and only if,

given that he can think that-one and can think that-two (where 'one' and 'two' represent propositional expressions), he can think that-one-and-two.

20. There is, however, another use of 'concept' in which one speaks of certain entities as concepts in a sense which, though acknowledging that they are items which thoughts can be "of" (and, indeed, uses this fact to pick them out), takes the relation between these entities and particular minds to be an "external" one. Thus the concept of triangularity would be an entity which would exist whether or not any particular mind had the concept of triangularity in the sense defined above. Concepts in this second sense are the sort of thing philosophers are talking about when they worry about "abstract entities" [5a] particularly about such intensional abstract entities as attributes and propositions. In this philosophically inspired sense one speaks of triangularity as a conceptual entity.[6]

21. Some philosophers (e.g. Plato, G. E. Moore and one of the many Lord Russells) have held that concepts in the second sense are "objective" in that their existence is independent – not only of this mind or that mind, but of mind *ueberhaupt*. Others, while stressing their independence of particular minds have stressed objectivity in the sense of inter-subjectivity (compare the objectivity of institutions), and have suggested that conceptual entities are mind-dependent in this broader sense.

IV

22. The position I wish to present and defend is, in its general outlines, a familiar one. I shall therefore stress those details which, as I see it, enable a reasonably straightforward resolution of the issues with which I began. Thus I shall assume that thinking at the distinctively human level and, in particular, the thinking involved in science, is essentially verbal activity. The verbal activity I have in mind is not the selection of words to convey to others, perhaps truthfully, perhaps deceitfully, an envisaged message concerning what one believes and/or intends; nor, in general, the use of words to perform illocutionary or perlocutionary actions. The linguistic activity I have in mind has, as its primary form, candid, spontaneous thinking-out-loud.[7] As such it rarely occurs, once we learn "to keep our thoughts to ourselves". Thus, most occurrent "verbal thinking" consists

of momentary propensities to think something "out loud", which propensities fail to erupt simply because of this general inhibition.

23. From this point of view, the classical conception of thoughts as pure occurrents is motivated by the familiar attempt to relate changes in dispositional properties to change in underlying non-dispositional states. The emptiness of the classical account of thought episodes is explained by the fact that it uses as its model for the description of the *intrinsic* nature of thought episodes (i.e. what they "consist of") concepts which, as we shall see, are largely functional in character.[8] Indeed, it is, by and large, the *non-functional* aspect of the model which are, save in their most generic aspects, discarded. After all, leaving aside functional considerations, thoughts *are* neurophysiological processes and this is an idea which no arm-chair philosophizing could turn into cash.

24. Now if thinking is, in the above sense, verbal activity, then ascribing a certain thought to a person by the use of "indirect discourse" rather than "direct quoting" is not simply analogous to, but identical with, telling what someone has said (or was disposed to say) by the use of indirect discourse instead of direct quotation.

25. The primary mode of being of the linguistic is in the linguistic activity of persons. The statement

Jones said '2+2=4'

where this simply has the force of Jones thought-out-loud '2+2=4' – i.e. where illocutionary force is absent – has the form

Jones '2+2=4'd

where "'2+2=4'd" stands to 'Jones' as verb to noun. It is because there is a range of verbal activities involving the uttering of '2+2=4' e.g. asserting, repeating, etc., that we give it the status of an adverb, and hence, in effect, require that even in the case of sheer thinking-out-loud there be a verb which it modifies. This is one source of the illusion that the concept of uttering '2+2=4' *assertively* (where the latter does not connote the illocutionary act of asserting) requires the neustic-phrastic distinction.

26. Now

(1)   Jones said '2+2=4'

is obviously not to be identified with

(2)   Jones uttered '2+2=4'

where this simply tells us that Jones produced sounds of the kind conventionally associated with the shapes of which *those* (the ones between the quotes) are samples. What is the difference? The answer clearly has *something* to do with "meaning". We are tempted to say that (1)=Jones uttered '2+2=4' *as meaning 2+2=4*. This is not incorrect, but also not illuminating.

27. To put our finger on what is involved, it will be useful to turn our attention away from language as activity to language as product, thus inscriptions, recordings and the like. If we can understand the meaning of "meaning" in the context, say, of inscriptions (external accusatives) we shall not be far from understanding what it means to speak of the meaning of utterances (internal accusatives).

28. Thus, consider the old chestnut

(3)   'Und' (in German) means *and*.

Two things are to be noted: (a) The subject of this sentence is a singular term. (b) The word with which it ends is an unusual use of the word 'and', for it is not serving as a sentential connective. Let me take up these two points in order.

29. Many philosophers have succumbed to the temptation to construe the subject of (3) as the name of a linguistic abstract entity, the German word 'und' as a universal which can (and does) have many instances. Yet this is a mistake which can (and does) cause irreparable damage. There are, indeed, many 'und's and they are, indeed, *instances* of a certain kind – 'und'-kind, we may call it. There are also many lions and they are instances of lion-kind. But it is important to distinguish between two singular terms which are in the neighborhood of the sortal predicate 'lion'. There is, in the first place, the singular term which belongs in the context

... is a non-empty class.

Ordinary language has no neat expression which does this job. 'The class of lions' will do. But there are also such terms as 'the lion' or 'a lion' or 'any lion':

> The lion (or a lion, or any lion) is tawny

where these are *roughly* equivalent in meaning to

> All lions are tawny.

Each of these expressions in its standard use in such sentences has "conversational implicatures", some of which are relevant to the linguistic examples which I shall shortly be giving. I call such singular terms "distributive singular terms" (DSTs).[9]

30. Thus the correct interpretation of the subject of (3) treats it not as an abstract singular term which designates an abstract entity, but as a distributive singular term. In other words (3) is, for our purposes, identical in sense with

(3¹)   The (or an, or any) 'und' (in German) means *and* or, equivalently, with

(3²)   'Und's (in German) mean *and*.

31. The second point to be noted about (3) was that it involved an atypical use of the word 'and', for it is clearly not functioning as a sentential connective. A natural move is to construe the context as a quoting one. This idea may tempt one to rewrite (3) as

(3³)   'Und' (in German) means 'and'

but quoting contexts are often such that to leave them unchanged while adding quotes to the quoted item changes the sense. And it is clear that (3) does not merely tell us that 'und' and 'and' have the same meaning, it in some sense *gives* the meaning. I have argued that the correct analysis of (3) is

(3⁴)   'Und's (in German) are ·and·s

where to be an ·and· is to be an item in any language which functions as 'and' does in our language. Roughly to say what an expression means is to classify it functionally by means of an illustrating sortal.[10]

32. According to this analysis, meaning is not a relation for the very simple reasons that 'means' is a specialized form of the copula.[11] Again, the meaning of an expression is its "use" (in the sense of function), in that to say what an expression means is to classify it by means of an illustrating functional sortal.

33. Notice that instead of "giving" the complex function of 'und' (in German) by using an illustrating functional sortal, we could, instead, have listed the syntactical rules which govern the word 'und' in the German language.[12]

34. The above, however, is but the entering wedge for the resolution of our problem. It provides the essential clues, but its significance is not yet manifest. For there are other ways of making meaning statements than by the use of 'means'. And it is these other ways which have generated much of the confusion and perplexity which are characteristic of the controversy over conceptual change.

35. Thus consider

(4)   'Dreieckig' (in German) stands for triangularity.

According to appearances (surface grammar) the following seem to be the case: (a) 'Triangularity' is a name. (b) It refers to a non-linguistic entity. (c) *Stands for* is a relation which, given the truth of (4), holds between a linguistic and a non-linguistic entity. I shall argue that (a), (b) and (c) merely *seem* to be the case, and that, contrary to the general opinion, to "countenance" statements like (4) is *not* to commit oneself to a Platonistic ontology.

36. The point grows directly out of our previous account of 'means' sentences. For there we encountered two ideas which can be put to good use: (a) 'Means' is a specialized form of the copula. (b) What follows 'means' is to be construed as a metalinguistic sortal. (c) The subject of a 'means' statement is a metalinguistic distributive singular term. To put these ideas to work all we need to do is to construe 'triangularity' as a metalinguistic distributive singular term, and 'stands for' as another (and more interesting) specialized copula.

37. Consider the following sentence, which is of a kind to which logicians have paid little attention

(5)   The pub is the poor man's club.

How are we to understand the copula 'is'? Only a most superficial reading would take (5) to be a statement of identity. Surely we have here a statement involving two distributive singular terms formed, respectively, from the sortals 'pub' and 'club'. It has the form

(6)   the K_1 is the ϕK_2

and is roughly equivalent to

(5^1)   Pubs are poor men's clubs.

38. I propose, therefore, that we read (4) as

(4^1)   The 'Dreieckig' is the German ·triangular·

which transforms into

(4^2)   'Dreieckig's are German ·triangular·s

or, which is the same thing,

(4^3)   'Dreieckig's (in German) are ·triangular·s.

39. According to this interpretation, (4) is simply another way of doing what is done by (3), i.e. giving a functional classification of certain inscriptions belonging to the German language. What is the point of having this second way? The answer is simply: Because *this* way of doing the job relates the classification to the truth context

(7)   Triangularity is true of a

which tells us, in first approximation, that

(8)   Expressions consisting of a ·triangular· appropriately concatenated with an ·a· are true.

40. In general, I suggest that so-called nominalizing devices which, when added to expressions, form the corresponding abstract singular terms,

thus '-ity', '-hood', '-ness', 'that...', etc., are to be construed as quoting contexts which (a) form metalinguistic functional sortals, and (b) turns them into distributive singular terms.

41. Thus 'triangularity' merely *looks* (to the eye bewitched by a certain picture) to be a name. It merely *looks* as though it referred to something non-linguistic. Applying to expressions in *any* language which do a certain job, its inter-linguistic reference is confused with a non-linguistic reference. Again 'stands for' merely *seems* to stand for a relation. It is, as 'means' proved to be, a specialized form of the copula.

V

42. Clearly the present occasion does not permit a systematic development of the semantical theory to which the preceding is but the preface. Yet it is not difficult to see its outlines, and enough has been said about it to prepare the way for its application to the problem with which we began.

43. Notice, for example, the new look of the problem of "identity conditions for attributes". Since talk about attributes is talk about linguistic "pieces", and not about Platonic objects, identity means sameness of functions, and belongs in a continuum with similarity of function.

44. Thus, after studying two games which use physically different materials and motions, we might decide that the two games are the "same", i.e. that we can find an abstract specification of correct and incorrect moves and positions such that it picks out for both games the moves and positions which are correct or incorrect according to their less abstractly formulated rules.

45. And by virtue of this fact we could say, for example, that the *Dame* of one game is the Queen of the other. By parity of reasoning, we can say that

f-ness $= g$-ness if and only if the rules for $\cdot f \cdot$ s are the same as the rules for $\cdot g \cdot$ s

46. One can also make sense of the idea that bishops are more like castles than they are like knights. Indeed, we are all accustomed to mak-

ing judgements of this kind. 'The bowler in cricket is like the pitcher in baseball.' We decide similarity of "pieces" with reference to the roles they are given by the rules.

47. I have often been asked, what does one gain by abandoning such standard Platonic entities as *triangularity* or *that 2+2=4* only to countenance such exotic abstract entities as functions, roles, rules and pieces. The answer is, of course, that the above strategy abandons nothing but a picture. Triangularity is not abandoned; rather 'triangularity' is *seen for what it is*, a metalinguistic distributive singular term.

48. And once the general point has been made that abstract singular terms are metalinguistic distributive singular terms, rather than labels of irreducible eternal objects, there is no reason why one should not use abstract singular terms and categories of abstract singular terms in explicating specific problems about language and meaning. For just as talk about triangularity can be unfolded into talk about ·triangular· inscriptions, so talk about any abstract entity can be unfolded into talk about linguistic or conceptual tokens.

VI

49. Let us now look at likeness of meaning from a somewhat different direction. Consider the familiar fact that isosceles triangularity and scalene triangularity are species of triangularity. In our framework this is spelled out as the fact that

·isosceles triangular·s

and

·scalene triangular·s

consists of a common predicate (a ·triangular·) concatenated with a modifier (an ·isosceles·, a ·scalene·) in such a way that ·triangular·, ·isosceles triangular·, and ·scalene triangular· constitute a fragment of a system of geometrical classification.[13]

50. The important point is that isosceles triangularity is to be construed as (isosceles triangular)-ity, the scope of the quoting context '-ity' being indicated by the parentheses. Contrast this with the contrast between

Euclidean triangularity and Riemannian triangularity. Here the scope of '-ity' is simply 'triangular'. Thus to talk about Euclidian triangularity is to talk not about

·Euclidian triangular·s
but about
Euclidian ·triangular·s

i.e. inscriptions which function as does our word 'triangular' when it is governed by specifically Euclidian principles.

51. Thus it is important to note that the use of the illustrating device to form functional sortals involves an important flexibility. Not all aspects of the functioning of the illustrating expression need be mobilized to serve as criteria for its application. Thus consider

Euclidian triangularity and Riemannian triangularity are varieties of triangularity.

This becomes

Euclidian ·triangular·s and Riemannian ·triangular·s are varieties of ·triangular·.

It is clear that the functioning of the illustrating word 'triangular' which is relevant to something's being a ·triangular· is a generic functioning which abstracts from the specific differences between Euclidian and Riemannian geometries.

52. Compare

Classical negation and intuitionistic negation are varieties of negation.

Here again the context makes clear just what aspects of the functioning of the illustrating term is being mobilized by the abstract singular term into which it is built. It is our intuitive appraisal of the functional similarity of expressions in different linguistic structures which grounds our willingness to make statements of this form.

53. Thus, suppose Jones opens a treatise on classical kinematics and

finds a sentence involving the function 'l', thus

(9)   ...l...

If Jones is a pre-relativity physicist he would be prepared to say

In (9), 'l' stands for length.

If, on the other hand, he is aware of relativity kinematics, he would be well advised to say

In (9) 'l' stands for *classical* length.

Yet it would be quite correct for him to say

In (9) 'l' stands for length

provided the context made it quite clear that this way was equivalent to

In (9) 'l' stands for *a length concept*.

54. We are now in a position to understand why philosophers are tempted to say that the change from classical kinematics to relativity kinematics is a change in belief about length, mass, velocity, simultaneity, etc. For consider the following schematic quotes from Newton-Jones and Einstein-Smith

N-J: ... l... v... m... simul...
E-S: ---l---v---m---simul---

Each statement expresses a belief. Each statement contains terms standing for length, velocity, mass, simultaneity, etc. Thus it is tempting to suppose that we are confronted by different beliefs involving *the* concepts of length, velocity, etc. But this is surely a mistake. We are confronted by different beliefs each of which involves *a* concept of length, velocity, simultaneity, etc.[14]

VII

55. The above account of meaning and likeness of meaning gives point to the idea of meaning change. Consider the evolution of a game. Suppose that at one point in the history of chess the piece which was checked and checkmated could capture like a knight as well as on adjacent squares.

Suppose that shortly thereafter, following a period of controversy, the community of chess-players decided that the game would be improved in certain respects if this power to capture like a knight were dropped. Would we not be willing to say not only that the game has changed, but that the king has changed? It is not as though a dog vanished and a cat took its place.

56. Then why not say that the concept of length changed during the transition from classical to relativity kinematics? After all, it did.

University of Pittsburgh.

NOTES

[1] The "observational" framework includes, of course, complex defined concepts which cannot themselves be perceptually applied.

[2] Sellars (1965).

[3] Even in such a case, however, one would be willing to pick out quantities in the two theories which correspond (more or less roughly) to the same empirical phenomena, thus 'loss of phlogiston' and 'oxidation' with combustion. Is it illuminating to say that 'loss of phlogiston' and 'oxidation' both stand for combustion in something like the sense in which 'l' in both special relativity and Newtonian kinematics stands for length? Obviously what is needed is reflection on what it is for a term to stand for a concept, a topic to which I shall soon turn.

[4] I say "with relative ease" because the fundamental issues in the conflict between "change of belief" and "change of concepts" theorists recur, though less dramatically, at the observation level. For an exploration of the connections between law-likeness and meaning see Sellars (1958), especially pp. 266ff.

[5] Of course this competition too would involve aspects of logical conflict in the tough sense, though at a different level.
Bergmann, G., 1955, 'Intentionality', *Semantica* Instituto di Studi Filosofiti, Rome.

[5a] Needless to say the concept of triangularity, as the ability to think thoughts to the effect that such and such an item is triangular, is also an abstract entity, though not the same one.

[6] Subtle and important distinctions are involved in the debate between those who prefer to use the abstract singular term and say that triangularity is a concept, and those who, like Frege, prefer to use the predicative expression and say, rather, that *triangular* is a concept. These issues lie outside the scope of my argument, though its implications for them are quite direct.

[7] See Sellars (1969a), also Chapter III of Sellars (1968b) and my unpublished Machette Lectures on 'The Structure of Knowledge'.

[8] See Sellars (1969b).

[9] See Sellars (1963a), reprinted as Chapter 5 in Sellars (1968a). Notice that I am not saying that all expressions of the form 'the K' which are not definite descriptions of an

individual *K* are DSTs. Thus in 'the lion once roamed the western plains', the subject is not a DST, for, though its sense is roughly equivalent to 'lions once roamed the western plains', it is not even remotely equivalent to 'all lions once roamed the western plains'. [See the papers of Ziff and Clark in this volume – eds.]

[10] It is important to note, for future reference, that it is an over-simplification to speak of "the" function of a certain expression in a given language. Various devices can be used to make it clear which functions of the word which is used to form the illustrating sortal are serving as criteria for its application.

[11] I am assuming, of course, without argument, that the copula 'is' does not stand for an "ontological *nexus*" (exemplification). The theory of predication is the crux of ontology. I have posed the issues in Sellars (1963b). Notice that from my point of view Bergmann is (mis) perceptive but consistent when he treats meaning as a nexus. See Bergmann (1955), reprinted in Bergmann (1960), pp. 3–38.

[12] In principle the rule governed uniformities which constitute a language (including our own) can be exhaustively described without the use of meaning statements, including those to be discussed below. In practice, the use of meaning statements (translation) is indispensable, for it provides a way of mobilizing our linguistic intuitions to classify expressions in terms of functions which we would find it difficult if not (practically) impossible to spell out in terms of explicit rules.

[13] The questions, 'what is a predicate? a predicate modifier? concatenation?' are of the greatest importance. On the present occasion, there is nothing to do but rely on intuitive considerations.

[14] There is also a sense in which N-J and E-S have different beliefs about length, etc., for, if they are engaged in controversy, we may say that they have different beliefs about *which* length, mass, velocity, etc. concepts *should be adopted*.

BIBLIOGRAPHY

Bergmann, G., 1955, 'Intentionality', *Semantica*, Instituto di Studi Filosofici, Rome.
Bergmann, G., 1960, *Meaning and Existence*, Madison.
Sellars, W., 1958, 'Counterfactuals, Dispositions and the Causal Modalities', in *Minnesota Studies in the Philosophy of Science*, Vol. II (ed. by H. Feigl, M. Scriven and G. Maxwell), Minneapolis, pp. 225–308.
Sellars, W., 1963a, 'Abstract Entities', *Review of Metaphysics* 16, 627–71.
Sellars, W., 1963b, *Science, Perception and Reality*, London.
Sellars, W., 1965, 'Scientific Realism and Irenic Instrumentalism', in *Boston Studies in the Philosophy of Science*, Vol. II (ed. by R. S. Cohen and M. W. Wartofsky), New York, pp. 171–204.
Sellars, W., 1968a, *Philosophical Perspectives*, Springfield, Ill.
Sellars, W., 1968b, *Science and Metaphysics*, London.
Sellars, W., 1969a, 'Language as Thought and as Communication', *Philosophy and Phenomenological Research* 30, 506–27.
Sellars, W., 1969b, 'Toward a Metaphysics of the Person', in *The Logical Way of Doing Things* (ed. by K. Lambert), New Haven, pp. 219–252.

KEITH LEHRER

EVIDENCE, MEANING AND CONCEPTUAL CHANGE: A SUBJECTIVE APPROACH*

Conceptual change is replete with semantic and epistemic implications. Fundamental changes in the way in which we conceive of the world formulated in laws and theories shift the meaning of words and the content of evidence. It is controversial whether conceptual change always shifts the meaning of terms in our laws and theories.[1] However, what statements we countenance as evidence may be altered whether or not conceptual change alters the meaning of words. Our main purpose in this paper is to formulate a theory of evidence explaining how conceptual change alters, not only what we infer from evidence, but the very content of evidence itself.

We shall argue that whether statements are selected as evidence depends on our estimates of the chances of statements being true and on which statements they compete with for selection. Our starting point will be subjective in the sense that our rules of evidence will be based on the subjective estimates and beliefs of men without assuming the truth of any other statements except those of logic. Any statement may be selected as evidence provided that the selection yields some positive gain in the value we may reasonably expect to obtain in our search for truth on the basis of our subjective estimates and beliefs. Finally, we shall consider how intersubjective agreement can restrain complete subjectivity of the individual in the rational selection of evidence.

I. EVIDENCE AND SUBJECTIVITY

We now turn to the problem of formulating a theory of evidence to explain how conceptual and semantic shifts influence the character of empirical evidence. Our starting point will be radically subjective, and this requires some defense.

Scientific inquiry is alleged to be objective because the rejection and acceptance of hypotheses and theories is based on objective empirical data. But this is a myth. What counts as data is revised as a result of conceptual change. Conceptual change is a change in the way in which we conceive of the world, that is, in our general beliefs about what the world is like. Hence our general beliefs shape the very content of our data and evidence. Belief, whether of a single man or a group of men, is a subjective factor. To explicate the implications of conceptual change for the revision of evidence, we shall build our theory of evidence on the base of subjectivity.

There is another reason for adopting a subjective approach to the theory of evidence. The other alternative is to make the selection of evidence a matter of apriori principles. Adamant empiricists, who insist that all evidence must consist of observation statements, are appealing to an apriori principle concerning evidence. Consider the requirement that only observation statements are evidence statements. What is the status of this requirement? It is not established by those observation statements allowed as evidence. Those statements cannot speak for themselves and lay claim to the status of evidence. It is postulated apriori by the empiricist.

The arbitrariness of such requirements, conventions, or postulates became clear historically when empiricists disagreed about what counts as an observation statement. Some said sense data statements were the evidence, others said that physical object statements were. Once the matter was settled, it became clear that the evidence statements philosophers fasten upon are precisely those that lead to the conclusions they and their associates believe. The ultimate basis for selecting evidence is subjective no matter how fervently we claim otherwise.

It might be thought that pragmatism provides an alternative to subjectivism. However, the rustic dress of pragmatism hardly conceals the natural shape of subjectivism that lies beneath. If what counts as evidence is to be decided in terms of how well that selection would serve our pragmatic ends, we immediately face the problem of deciding how likely various selections are to serve those ends. To decide this, we must antecedently assume certain statements to be evidence which we cannot do in the selection of evidence, or we must rely upon our subjective beliefs and estimates. There is no other alternative.

II. SUBJECTIVE PROBABILITY

With these preliminary defenses of the subjective way, let us now consider how to construct a theory of evidence from a subjective base. What sort of subjective base may we employ? We shall assume the viability of a concept of subjective probability and formulate a theory of evidence based on it. Subsequently, we shall examine a variation of the theory based on subjective conviction and intersubjective agreement as well as subjective probabilities.

Subjective probabilities have been construed as degrees of belief in contrast to such objective probabilities as frequencies or propensities. The degree of belief that a man has in a statement is interpreted behavioristically in terms of what the man considers to be a fair betting quotient for the truth of a statment. He assigns a subjective probability of m/n to a statement s if and only if he would consider a bet on the truth of s to be fair if it paid n dollars on a bet of m dollars. Subjective probability is thus determined by the betting preferences of the subject. The subjective conception of probability has been advanced by DeFinetti, Savage, Jeffrey and others.[2] I shall take for granted the viability of assigning subjective probabilities to statements a man might consider taking as evidence.

It is required that such subjective probabilities satisfy the calculus in order to be considered rational. Subjective probabilities that fail to meet this requirement are called *incoherent*. One justification for requiring this is that betting quotients that are incoherent allow dutch book to be made against the subject. That is, if a subject accepts a set of betting quotients that do not conform to the calculus of probability and offers to cover bets made at those odds, then a sufficiently skilled calculator can make bets in such a way that the subject is certain to lose no matter what statements turn out to be true or false. Thus, the requirement that subjective probabilities conform to the calculus of probability is imposed to insure against certain loss.

III. BETTING WITH NATURE

This line of argument has been criticized by Putnam.[3] He argues that there is no reason to suppose that nature is a skilled calculator who will make dutch book against you whenever your betting quotients fail to conform to the calculus of probability. Indeed, if your opponent is not such a skilled calculator, and takes the wrong bets, then he may place himself in a position where *he* is certain to lose because the betting quotients do not conform to the calculus. Thus, a man whose subjective probabilities fail to conform to the calculus of probability and who is guided by those probabilities in action and belief may not be any more unsuccessful or incorrect then a man whose subjective probabilities conform to the calculus. Nature may never catch the odd man out.

One way to meet Putnam's objection is to argue that the *ideally* rational man should not be a *certain* loser even if nature is an ideally skilled calculator. If Putman sees noting ideal in such a man, we have reached an impasse. There is, however, another way of meeting the objection. First, we abandon the standard treatment of subjective probabilities as degrees of belief. There are three reasons for doing so. First, there is no reason to suppose that subjective probabilities reflect the felt intensity of conviction of a man in the truth of a statement. He may feel strongly convinced of the truth of a statement but offer stronger odds in favor of the truth of some other statement of whose truth he feels less convinced. This may occur if the man trusts the opinion of another more strongly than his own conviction. Hence the identification of betting quotients with degrees of belief is phenomenologically unsound.

Second, as Levi has pointed out, if we identify degrees of belief with betting quotients, then we confront the problem of having to explicate the relation between degrees of belief and belief simpliciter.[4] If a man believes something to any degree greater than zero, than, according to Levi, he believes it. But a man may not believe a statement even though he would agree that there is one chance in a thousand that the statement is true. If we reject Levi's claim, we face the problem of saying at what degree of belief a man may be said to believe a statement to be true. One could repudiate this problem by claiming that the concept of belief simpliciter is philosophically or scientifically untenable. However, philolosophers adopting this position usually have no argument against the concept of belief simpliciter other than that they wish to identify degrees of belief with betting quotients and reject belief simpliciter as a prescientific relic of common sense. But there does seem to be such a thing as belief. Men are convinced they believe some things and disbelieve others. It is worth investigating whether there is some treatment of belief and subjective probability that better fits the psychological phenomena familiar to us all.

IV. ESTIMATING TRUTH

We propose to equate subjective probabilities, not with degrees of belief, but, as Carnap proposed, with a subjective estimate of the objective probability of the statement being true, that is, with an estimate of a frequency or propensity.[5] We continue to equate subjective probability with betting quotients as suggested above. The betting quotient and subjective probability will not be treated as degrees of belief in the truth of a statement but as estimates by the subject of the frequency with which such statements are true. This explains why a man might not feel convinced of the truth of a statement even though he assigns a high subjective probability to it. He might not feel strongly convinced of the truth of a statement even though he estimates that the chances of the statements being true are very good. This might result because the man is relying on the authority of others rather than on first hand experience. What one sees with one's own eyes often produces strong convictions, but reason and past experience keep us from betting our savings on them.

Moreover, a subjective estimate of the objective probability of a statement as m/n may be equated with the belief that the probability is m/n. Hence, we may equate subjective probabilities, not with degrees of belief in the truth of a statement, but with a belief concerning the objective probability of the statement being true. This equation provides an answer to Putnam's objection. Consider the frequency theory of probability in which probabilities are interpreted as limits of relative frequencies. As frequentists have shown, Reichenbach for example, once probabilities are equated with the limits of relative frequencies, then the calculus of probability may be deduced from the mathematical theory of limits.[6] In short, if probabilities are limits of relative frequencies, then it is a mathematical truth that such probabilities satisfy the calculus. Hence if subjective probabilities are interpreted as estimates of objective probabilities, limits of relative frequencies, then either subjective probabilities must satisfy the calculus or they will be demonstrably incorrect.

The justification for requiring, as a condition of ideal rationality, that the subjective probabilities of a man conform to the calculus of probability is the same as the justification for requiring that a man be logically consistent. The violation of either requirement brings with it the certainty of error. If a man is logically inconsistent in what he affirms or believes, then he cannot possibly be correct in what he believes or affirms. Similarly, if the subjective estimates of a man concerning objective probabilities are incoherent, then he cannot possibly be correct in what he estimates. Of course, some of the beliefs of a man whose beliefs are inconsistent with each other may be correct, and some of the estimates of a man whose estimates are incoherent may be correct. But error is inevitable when beliefs are inconsistent and estimates are incoherent. Therefore, subjective probabilities must conform to the calculus of probability, not simply because dutch book can be made against one otherwise, but because subjective probabilities are estimates of ratios which, as a sheer matter of mathematics, conform to that calculus.

Thus, our selection of evidence statements will be based on subjective probabilities which are estimates of objective probabilities satisfying the calculus of probabilities. This basis for selection is easily justified. The selection of evidence statements should be guided by the search for truth, hence, our selection of evidence should be based on our estimate of the chances of statements being true. Subjective probabilities are such estimates.

Subjective probabilities as thus construed may be influenced by any factors whatever. Our desire to explain, to obtain simplicity, to achieve consensus, to be guided by our senses, may all shape our estimate of the chances of a statement being true. Conceptual change is especially effective in shifting our subjective probabilities. A change in the meaning of a statement may radically alter the assignment of subjective probability. Hence, the selection of evidence based on subjective probabilities will be sensitive to the factors producing changes in the way in which we conceive the world and in the meaning of the statements we employ to express those conceptions. We shall formulate a rule of evidence based on subjective probabilities enabling us to explicate the precise way in which the selection of evidence is shaped by conceptual change.

V. A RULE OF EXPECTED VALUE

To formulate a rule for the rational selection of evidence, we employ a probability functor marked relative to time. Subjective probabilities are relative to both a person and a time. Men may differ in their subjective probabilities, and the same man may change from one time to another.

We suppress references to persons for notational simplicity. We assume that for all the statements of some language L, there is a subjective probability, $p_j(h)$, which is the subjective probability of h at time j for the person in question.

The rational selection of evidence should be guided by the search for truth. The search for truth has two quite different dimensions. One is to avoid error in what one selects as evidence. The other is to select statements that are informative in terms of their empirical content.[7] A decision theoretic approach to the problem tells us that it is reasonable to accept a statement as evidence if and only if the statement has a maximum of expected value. The expected value of selecting a statement as evidence is a function of the value we attach to possible outcomes of such a selection and the subjective probabilities we assign to those outcomes. In the search for truth, there are two relevant outcomes, truth and error. Following Hempel, Levi, Hintikka, and others, we introduce a pair of value functors, $VT_j(h)$ and $VF_j(h)$, symbolizing the value at time j we attach, in our search for truth, to the outcomes of selecting h as evidence when h is true and selecting h as evidence when h is false, respectively.[8] We obtain the following equation for the expected value of selecting h as evidence at time j, symbolized, $EV_j(h)$:

$$EV_j(h) = p_j(h)VT_j(h) + p_j(-h)VF_j(h).$$

To obtain a rule for the selection of evidence we must provide a method for ascertaining the quantities to be assigned to the value functors.

We have said that the search for truth involves the two objectives of avoiding error in what one selects as evidence and selecting statements that are informative in terms of empirical content. Since the probabilities express our estimate of the chances of avoiding error, the other factor in the above equation, the value functor, may plausibly be assigned the role of expressing our estimate of content. A number of authors propose that we accept the following equalities or some variation there of:[9]

$$VT_j(h) = 1 - \text{cont}_j(-h)$$
$$VF_j(h) = -\text{cont}_j(-h).$$

If we are seeking to avoid error and we select h as evidence when h is false, then our loss is equal to whatever the denial of h tells us, to the content of $-h$. On the other hand, if we select h as evidence when h is true, then our gain is equal to what value remains after subtracting from unity what we would have lost had *h* been false.

VI. MEASURING CONTENT

This leaves us with the problem of how are we to measure content, that is, assign values to the function, $\text{cont}_j(h)$. Most philosophers choose to measure content in terms of what the statement excludes by logical contradiction. We might call this conception of content the *logical* or *semantic* conception.[10] Though this is an important and useful measure of content, it is not the one that is appropriate in this context. The reason is that the selection of a statement as evidence excludes other statements from being selected even though it does not exclude them by logical contradiction. The selection of one statement as evidence may count against the selection of another even though the two statements are not logically contradictory. We shall seek to measure the content of a statement in terms of what other statements it excludes from being selected as evidence whether or not it contradicts those statements.

Feyerabend has argued that the empirical content of a statement depends not only on its logical relations to other statements but also on which statements it competes with. The empirical content of a statement is determined by the statements with which it conflicts.[11] It is a natural extension of this idea, one we shall adopt, that the empirical content of the denial of a statement is determined by the probability of a most probable statement with which it competes. What we have lost, when we select a false statement as evidence, will be measured in terms of its most probable competitor. The loss in value when we select a false statement as evidence is equal to the probability of some most probable statement with which that false statement competes. The value we gain when we select a true statement as evidence is equal to what remains when we subtract the probability of some most probable competitor from the maximum value it is possible to gain.

We need a definition of competition to complete our analysis of content. Obviously, logically inconsistent statements compete with each other. But statements may compete with each other even though they are mutually consistent. For, there are many statements which, though logically consistent, conflict empirically in the sense that they cannot both be true as a matter of empirical fact. Moreover, when we are selecting evidence statements we have no way of deciding which pairs of logically consistent statements conflict in this way. To decide whether a pair of statements conflict empirically, we need to appeal to evidence.

Faced with the problem of selecting evidence when we have no way of telling whether statements conflict empirically, how are we to decide which statements compete with each other? We propose a man assume two statments compete if it is logically possible that the pair of statements conflict empirically, that is, cannot both be true as a matter of empirical fact. By so doing, he will avoid selecting a statement as evidence that conflicts empirically with some other statement he has selected as evidence. Assuming that a statement competes with every other statement with which it could possibly conflict empirically is necessary to escape from internal conflicts among the statements we select as our evidence.

VII. A DEFINITION OF COMPETITION

Under what conditions is it logically impossible for a pair of statements to conflict empirically? Consider any two statements, each logically consistent, such that one logically implies the other. Such a pair of statements cannot conflict empirically, because the truth of one logically implies the truth of the other. If *each* of the statements can be true, then they can both be true. On the other hand, if two statements are such that neither logically implies the other, then it is logically possible that, as a matter of fact, though each of them can be true, they cannot both be true. Consequently, we assume that two statements compete with each other if and only if neither of them logically implies the other. The following is our definition of competition:

Comp. h and k compete if and only if neither $h \vdash k$ nor $k \vdash h$.

From this definition we can obtain a definition of content.

VIII. A DEFINITION OF CONTENT

We said above that the content of the denial of a statement is to be identified with the probability of some most probable statement with which the statement competes. Let us adopt the convention of letting h^* stand for some arbitrary most probable statement with which h competes. We define content as follows:

$$\text{cont}_j(\text{-}h) = p_j(h^*).$$

This definition of content runs contrary to a good deal of recent work. We shall not undertake an examination or defense of the general properties of this conception of content but restrict ourselves to considering the consequences of applying this conception to the selection of evidence statements. What is important for our purposes are the following equalities which result from the earlier equations of value and content:

$$VT_j(h) = 1 - p_j(h^*)$$
$$VF_j(h) = -p_j(h^*)$$

What we lose when we select a false statement as evidence is equal to our estimate of the chance of being right had we selected some strongest competitor for evidence instead of that false statement. What we gain when we select a true statement as evidence is equal to what value remains when we substract from unity the value we would have lost had the statement been false. Unity represents the maximum value one can obtain by selecting a statement as evidence.

IX. STRONGEST COMPETITORS

We can find a method for constructing a strongest competitor for any given sentence in finite languages, that is, languages having a finite number of individual constants and predicates. The number of primitive expressions may be astronomically large, and the number of sentences generated from such resources by truth functional manipulation will be infinite. However, there will be some finite set of sentences of the language such that every sentence of the language will be logically equivalent to some member of that finite set. Carnap's state descriptions provide the basis for the construction of such a set.[12] Every sentence of a finite language is logically equivalent to either a state description, a disjunction of statement descriptions, or a conjunction of two state descriptions. The set of state descriptions constitute a *basic partition* of statements, that is, a set of statements that are logically inconsistent in pairs and logically exhaustive as a set. Every contingent statement of the language is logically equivalent to a state description or a disjunction of state descriptions.

However, this method of specifying a finite language and basic partition is not the only alternative. We may start with a basic partition and define a language as the set of truth functional combinations of members of the partition. One may obtain finite languages by the second method that we cannot obtain by the first. For example, we can have sentences containing quantifiers, or sentences containing quantified sentences as conjuncts, as members of the basic partition. There are finite languages, ones containing no individual constants, that tell us exactly what the individuals of the universe are like, those that would result by taking Hintikka's Q-constituents as members of our basic partition.[13] For this reason, we shall use the second method, starting with a basic partition, to specify languages.

We shall be concerned with the selection of evidence statements in finite languages specified by this method. Such a restriction is not as limiting as it might appear. As we have argued elsewhere, when confronted by some choice of statements as evidence, the choice may be interpreted in finite terms even when it pertains to quantitative matter in which there are theoretically an infinite number of options.[14] For example, if the choice of an evidence statement concerns a statement specifying the length of some object, we must admit that there are theoretically an infinite, indeed, nondenumerably infinite number of choices. But in practise we shall be limited to measuring devices having a least measure. Consequently, our choice will actually be confined to a finite number of options. Thus, the finite character of a language does not preclude the consideration and selection of quantitative statements as evidence. We shall thus assume a basic partition P of sentences $m_1, m_2, \ldots m_n$ and a language L, which may be a finite part of some infinite language, where L consists of members of the partition and truth functional combinations of such members. The vocabulary of L will be restricted to whatever predicates and individual constants are contained in the members of P, but the interpretation of those terms as well as the specification of the rules of logic may be given in the larger language of which L is but a part.

We can provide a method for finding a strongest competitor in L for any given statment in language L. No contradiction or logical truth competes with any statement, because contradictions have all statements as llogical consequences, and logical truths are logical consequences of al statements.[15] Those statements which are neither contradictions nor logical truths, contingent statements, are logically equivalent to either some member of the partition P or to some disjunction of members of P in alphabetical order, each disjunct being a different member. Suppose s in L is logically equivalent to a member. The most probable statement in L that is neither logically implied by s nor logically implies s will be the disjunction of all the remaining members of P, because probability is additive. If s is logically equivalent to the requisite form of disjunction d of two or more members, then we may obtain a strongest competitor for s by disjoining all members of P that are not disjuncts of d to all members of P that are disjuncts of d save one, that one being a disjunct of lowest probability. The resultant disjunction will be some most probable statement that neither logically implies nor is logically implied by s.

We obtain a more formal statement of this method for finding a strongest competitor for h as follows:

$p_j(h^*) = p_j(d_h)$, where d_h competes with h and is a disjunction of all members of P in alphabetical order except one member, m_i, such that for any member m_k of P, if $m_k \vdash h$, then $p_j(m_i) \leqslant p_j(m_k)$.

This method provides us with a measure of empirical content for every contingent statement in L. No assignment is made to contradictions or logical truths as is appropriate for a measure of *empirical* content.

X. CHOOSING POSITIVE GAIN

With the foregoing definitions of content, of competition, and strongest competitor, let us examine the implications for the expected value of selecting a statement as evidence. Our equation for expected value becomes the following:

$$EV_j(h) = p_j(h)p_j(\text{-}h^*) + p_j(\text{-}h)(-p_j(h^*))$$
$$EV_j(h) = p_j(h)p_j(\text{-}h^*) - p_j(\text{-}h)p_j(h^*)$$
$$EV_j(h) = p_j(h) - p_j(h^*).$$

The latter equality enables us to calculate the expected utility of selecting a statement as evidence from the subjective probability of the statement and its strongest competitor. Obviously, the expected value of selecting a statement will be positive if and only if the subjective probability of the statement is greater than the subjective probability of its strongest competitor.

The usual decision theoretic rule is to maximize expected value. However, there are special conditions surrounding our application of this theory which necessitate some modification. First, it is possible that no statement in L will have an expected value other than zero. This will result when all members of the ultimate partition have the same probability and hence every statement will have the same probability as its strongest competitor. If we were to select as evidence those statements that maximize expected value, we could select any contingent statement because all would have an expected utility of zero. This is obviously not satisfactory as a rule for selecting evidence. We avoid this consequence by requiring that the expected value of selecting a statement as evidence be positive. We could then adopt the principle affirming that it is reasonable to select a statement as evidence if the expected value of the selection is maximal and positive.

We need not, however, restrict our selection of evidence statements to those having a maximal expected value. The customary application of rules of expected utility is to a situation in which an agent is faced with a choice between mutually exclusive courses of action. Since only one action can be chosen, it is only reasonable to choose that action which has a maximum of expected value. When we are selecting evidence statements we are not faced with a situation in which all alternatives are mutually exclusive. We can select more than one statement as evidence.

Any statement selected as evidence must have a *positive* expected value to insure that the statement selected competes favorably with its strongest competitor. Moreover, any statement that is more probable than its strongest competitor is such that the expected value of selecting it as evidence is positive. Since we are not limited to selecting one statement as evidence, it is reasonable to select all those statements as evidence when the expected value of such a selection is positive. When decision theory is applied to action we have to forego the positive expected value of choosing some actions in order to chose that action whose expected value is maximal. But there is no need to sacrifice the positive expected value of selecting one statement as evidence in order to obtain the maximal expected value of selecting another. Instead, it is reasonable to obtain all positive expected value by selecting as evidence, not only that statement whose expected value is positive and maximal, but also all those statements whose selection gives us positive expected value.

XI. A RULE OF EVIDENCE

Thus we propose to adopt the following rule for selecting statements as evidence, where we let '$E_j(e)$' mean 'e is evidence at time j':

RE. $E_j(e)$ if and only if $EV_j(e) > 0$.

The rule is equivalent to the following rule:

$E_j(e)$ if and only if it is not the case that $e \vdash \ulcorner p \,\&\, \text{-}p \urcorner$ or that $\vdash e$, and for any s such that e competes with s, $p_j(e) > p_j(s)$.

This rule has been shown to satisfy the following conditions:[16]

E1. If $\vdash \ulcorner e \equiv e' \urcorner$, then $E_j(e)$ if and only if $E_j(e')$.
E2. If $E_j(e)$ and $E_j(e')$, then $E_j(e \,\&\, e')$.
E3. There is some statement T such that $E_j(T)$ and for any statement s, if $E_j(s)$, then $T \vdash s$.

This says that there is a statement of total evidence selected as evidence.

E4. A statement T satisfying E3 is such that it is not the case that $T \vdash p$, for some p.

This condition says that the statement of total evidence is logically consistent and, hence, that the set of evidence statements selected by RE is logically consistent.

Another important feature of RE is that it guarantees that any true statement selected as evidence will be more probable than any statement except its logical consequences, and, consequently, will be more probable than any false statement. Thus, if the subjective probabilities are good estimates of frequencies or propensities, then an evidence statement, if true, is such that its truth is explained by those probabilities. If, contrary to RE, a true statement selected as evidence could be less probable, or even equal in probability to some false statement, then the probabilities would not explain the truth of the evidence statement.

In addition to these general characteristics of the rule of evidence we have formulated, it is possible to give an exact catalogue of the results obtained from the rule. A member of the partition P is selected as evidence if and only if it has a probability greater than its denial, that is, greater than $\frac{1}{2}$. A contingent disjunction of members of the partition P is selected as evidence if and only if the disjunction is more probable than any other such disjunction which it does not logically imply. Only contingent statements are selected, and all contingent statements are logically equivalent to members or such disjunctions. These considerations have the consequence that if some member of the partition has a lower probability than any other member, then the disjunction of all other members is selected as evidence. Moreover, if a disjunction of members is such that the least probable disjunct is more probable than the disjunction of all members not included in the disjunction, then again that disjunction shall be selected as evidence. Moreover, such disjunctions are the only disjunctions selected as evidence.[17]

XII. EVIDENCE AND DOXASTIC SYSTEMS

As it stands, rule RE is likely to favor belief statements as evidence because a person is likely to estimate that such statements have a very good chance of being correct. For example, consider the statement that I see a pen on the table. This statement is neither logically implied by nor does it logically imply the belief statement that I *believe* I see a pen on the table. I may see a pen on the table without believing that I do and I may believe that I see a pen on the table when I do not. Consequently, the statement that I see a pen on the table competes with the statement that I believe that I see a pen on the table. It would be natural for me to estimate that the latter statement has a better chance of being true than the former, that is, to assign a higher subjective probability to the belief statement. The reason is that the belief statement seems safer. Even if I am wrong about the claim that I see a pen, I may be right about the claim that I believe I see it.

The foregoing result may be avoided by modification of our rule of evidence. So far we have conceived of a man as selecting evidence with only his subjective probabilities to guide him. But it is consistent with our subjective approach to allow a man to be guided by other subjective considerations as well. Indeed, whatever a man believes, the statement that he believes it may reasonably determine his selection of evidence. If a man believes that p, it does not by any means follow that it would be reasonable for him to automatically include the statement that p in his evidence. But it would be reasonable for him to make his selection of evidence conditional on the statement that he *believes* that p.

Such a procedure would, of course, only be suitable for those men who regard belief statements as safe and hence assign a very high subjective probability to them. There are some philosophers who profess that there are no such things as beliefs or other mental entities. We may perhaps retort, with an unsympathetic grin, that if such philosophers do not believe what they say, then we may ignore their claim as insincere, and if they do believe what they say, then we may ignore their claim as incorrect. Thus, we ignore the objection.

Our proposal for conditionalizing evidence selection to belief may be rendered precise. Assume that a man A has a set of beliefs at time j. The set of statements of the form A believes that p, A believes that q, and so forth describing all his beliefs at j we shall call his doxastic system at j.[18] We then assume that there is at least one statement that logically implies all and only those statements logically implied by a doxastic system at time j. We symbolize that statement as d_j. We now make our definitions and equations relative to d_j.

XIII. RELATIVIZING TO DOXASTIC SYSTEMS

We let "$p_j(h, d_j)$" mean "the probability of h relative to d_j at j", "$EV_j(h, d_j)$" mean "the expected value of h relative to d_j at j", "$E_j(e, d_j)$" mean "e is evidence relative to d_j at j", and so forth for the other relativized functions. Next we define competition relative to d_j. One statement competes with a second relative to d_j if and only if the first does not logically imply the second relative to d_j. Formally, the relativized definition of competition is a follows:

> h competes with k relative to d_j if and only if it is not the case that $\ulcorner h \& d_j \urcorner \vdash k$ or that $\ulcorner k \& d_j \urcorner \vdash h$.

To define the notion of a strongest competitor relative to d_j we shall appeal to a subset of the basic partition P, to what Levi has called the basic partition truncated with respect to d_j.[19] The truncated partition, P_{d_j}, consists of those members of P consistent with d_j. We thus obtain the following definitions and equalities:

$$EV_j(h, d_j) = p_j(h, d_j) \, VT_j(h, d_j) + p_j(-h, d_j) \, VF_j(h, d_j)$$
$$VT_j(h, d_j) = 1 - p_j(h^*, d_j)$$
$$VF_j(h, d_j) = -p_j(h^*, d_j)$$
$$p_j(h^*, d_j) = p_j(s), \text{ where } s \text{ competes with } h \text{ relative to } d_j \text{ and is}$$
a disjunction of all members of P_{d_j} except one member m_k, such that for any member m_i of P_{d_j}, if $m_i \vdash h$, then
$$p_j(m_k, d_j) \leqslant p_j(m_i, d_j).$$

From these definitions we find that statements contradicted by or logically implied by d_j are not assigned any measure of epistemic value relative to d_j; just as contradictions and logical truths were not assigned any such measure earlier.

From the above, we deduce the following equality

$$EV_j(h, d_j) = p_j(h, d_j) - p_j(h^*, d_j)$$

and adopt the rule

RED. $E_j(e, d_j)$ if and only if $EV_j(e, d_j) > 0$.

The results of applying rule RED for the selection of evidence relative to a doxastic system are exactly parallel to the results of applying rule RE. The rule is equivalent to the following:

$E_j(e, d_j)$ if and only if it is not the case that $\ulcorner e \, \& \, d_j \urcorner \vdash \ulcorner p \, \& \, \text{-} p \urcorner$ or that $d_j \vdash e$, and for any s such that e competes with s relative to d_j, $p_j(e, d_j) > p_j(s, d_j)$.

The rule satisfies the following conditions:

E1.   If $d_j \vdash \ulcorner e \equiv e' \urcorner$, then $E_j(e, d_j)$ if and only if $E_j(e', d_j)$.
E2.   If $E_j(e, d_j)$ and $E_j(e', d_j)$ then $E_j(e \, \& \, e', d_j)$
E3.   There is some statement T such that $E_j(T, d_j)$ and for any statement s, if $E_j(s, d_j)$, then $\ulcorner d_j \, \& \, T \urcorner \vdash s$.
E4.   A statement T satisfying E3 is such that it is not the case that $\ulcorner d_j \, \& \, T \urcorner \vdash p$, for some p.

A member of the truncated partition is selected as evidence if and only if it is more probable on d_j than its denial. A disjunction of members of the truncated partition not logically implied by d_j is selected as evidence if and only if the disjunction is more probable on d_j than any other such disjunction not logically implied by the first in conjunction with d_j. Nothing logically implied by or contradicted by d_j is selected as evidence.

The close parallel between RE and RED may suggest that one is an unnecessary duplication of the other. This is not so for two reasons. As we have noted, some philosophers wish to deny the existence of beliefs altogether. Though we have rejected this position, we note that rule RE could be adopted consistently with such a philosophy provided that subjective probabilities were interpreted as betting propensities but not beliefs. Hence rule RE but not rule RED is available to such a philosopher. Moreover, when we consider the implications of conceptual change for the selection of evidence statements, we shall discover that rule RED provides an additional basis for understanding the influence of conceptual and theoretical factors on observation and perception.

XIV. EVIDENCE AND CONCEPTUAL CHANGE

We now turn to consider the way in which conceptual change influences the selection of evidence by rules RE and RED. On either rule, the selection of evidence is sensitive to shifts in subjective probabilities. Subjective probabilities may shift in response to a multiplicity of forces. Whatever influences men as they try to shape a conception of the universe satisfying their desires for coherence, simplicity, completeness and so forth may alter their subjective probabilities. Thus, a statement that is more probable than its strongest competitor at one time may subsequently undergo a downward turn in probability and no longer compete favorably. It will then be rejected as evidence. It has been noted by philosophers of a pragmatic bent that evidence statements can lose their evidential status, but they have not explained in detail how this transpires. On rules RE and RED, the ejection of statements from evidence can result from a simple shift in probabilities rendering a statement less probable than its competitors. Conversely, a statement not selected as evidence at one time may subsequently be selected as evidence because it has become more probable than its competitors. The shifting of statements in and out of the office of evidence is characteristic of both rules.

We noted above that conceptual changes are usually understood as changes of fundamental laws and theoretical principles. It is controversial whether such changes always involve changes in the meaning of words. We can better understand the implications of this controversy by appeal to rules RE and RED. Suppose that L is a statement which was not formerly thought to be a law and has now shifted to that status. If we suppose that this results in a shift in the meaning of words so that L is now true by the meaning of words whereas formerly it was not, then L must be assigned a probability of unity. However, it is also true that no statement will then compete with L because L will be logically implied by any statement, as are all truths of logic and semantics. Hence, no statement will need to compete with L to be selected as evidence.

On the other hand, if we suppose the meaning of L is unaltered as a result of becoming a law and L is *not* true by the meaning of words, then other statements must compete with L if they do not logically imply L according to RE, or if they do not logically imply L in conjunction with d_j according to RED. Hence, the effect of treating fundamental laws and theoretical principles as true by virtue of the meaning of words is that by so doing one clears the way for other statements to be selected as evidence. Someone who favors a very liberal theory concerning what is to be admitted as evidence may thus find some reason for embracing such a semantic theory.

XV. PROBABILITY SHIFTS

Subjective probabilities can shift from one time to another in any manner whatsoever provided that they conform to the calculus of probability at each time. Thus, it should not be assumed that the selection of a statement as evidence is equivalent to shifting the probability of the statement to one.[20] For finite languages, we adopt the condition that $p_j(h, e) = 1$ only if $e \vdash h$ and $p_j(h) = 1$ only if $\vdash h$. To allow that a statement be selected as evidence without being a logical truth, we reject the proposal that a statement automatically acquires a probability of 1 when selected as evidence.

The selection of a statement as evidence for the purposes of inductive inference and scientific inquiry should not be expected to produce any shift in the probability of the statement. When a man selects a statement as evidence because it competes favorably with other statements for that status, he need not have any reason for shifting his estimate of the chances of the statement being true or for changing his betting preferences. The selection of a statement as evidence does not afford a man any reason for assigning a different subjective probability to the statement.

Having noted in general terms that the selection of evidence need not shift probabilities and that conceptual change may shift the probabilities, let us consider some specific ways in which conceptual change may shift probabilities. Suppose that some conceptual change is embodied in a statement G which is a law or theoretical principle. In what way does this change shift the probabilities of other statements? If the shift renders G true by the meaning of words, then, letting p_0 be the earlier probability function and p_1 the probability function subsequent to the shift, we obtain the following formula for determining the probability p_1 of any statement h on the basis of p_0:

$$p_1(h) = p_0(h, G).$$

On the other hand, if the conceptual change alters the probability of G without any change in the meaning of words and G remains a contingent statement having a probability of neither 1 nor 0, then the following formula, taken from Jeffrey, may be employed:

$$p_1(h) = p_0(h, G) p_1(G) + p_0(h, -G) p_1(-G)$$

provided that

$$p_1(h, G) = p_0(h, G) \quad \text{and} \quad p_1(h, -G) = p_0(h, -G).^{21}$$

If the conditional probabilities remain constant, we may, once we assign a probability to G at time 1 subsequent to conceptual change, calculate the probabilities of other statements as well. Thus, given these conditions, we have a method for explicating the way in which a shift in the subjective probability of statement G produced by conceptual change is propagated throughout the language. Of course, we have no guarantee of conditional probabilities remaining constant so that we can employ the formula above to calculate the shift in probabilities resulting from conceptual change. It is nevertheless an illustration of one way in which probabilities may shift as a consequence of conceptual changes that do not alter the meaning of words. The shift in subjective probability thus produced may alter what is selected as evidence by both RE and RED.

The formula employed to explicate the way in which subjective probabilities shift when some contingent statement changes meaning and becomes true by the meaning of words may also be used to explain how probabilities shift when some contingent statement changes meaning and becomes false by the meaning of words. When we wish to calculate the shift in the probabilities of statements resulting from some statement F becoming false by the meaning of words, we may employ the following formula:

$$p_1(h) = p_0(h, -F).$$

However, when conceptual change renders a statement contingent which was not so formerly, the explication of probability shifts is somewhat more complicated.

Suppose that we have a partition P which is a set of statements that are logically disjoint in pairs and exhaustive as a set. Such a partition may contain a member that is a contradiction. A contradiction is inconsistent with any statement and a set of statements that is logically exhaustive remains so when a contradiction is added to the set. Let us call partitions that contain logically inconsistent members *inflated* partitions. An inflated partition may contain a contradictory member which conjoins two sentences S1 and S2 such that it is true by the meaning of words, by a meaning postulate of the language, that -(S1 & S2). Now suppose that because of conceptual change the latter statement is no longer true by the meaning of words: it is no longer a meaning postulate that -(S1 & S2). In that case, those members of the partition that were inflationary solely because they contained both S1 and S2 will be inflationary no longer. They now will be contingent statements which may compete with other statements for selection as evidence. Some value will be attached to their selection.

When an inflationary member of a partition becomes contingent and noninflationary through conceptual change, some positive probability must be assigned to that member at the expense of the probabilities of other members. This may occur in a variety of ways, but there is one simple and natural redistribution of probabilities among other members of the partition. If at time 0 there are m noninflationary members of P and p_n is an inflationary member that has become noninflationary at time 1 through conceptual change, then any noninflationary member p_j at time 0 might be expected to lose an equal share of probability to supply the cositive probability of p_n at time 1 subsequent to conceptual change. We opuld then calculate the probability of any such p_j at time 1 according to the following formula:

$$p_1(p_j) = p_0(p_j) - p_1(p_n)/m.$$

The probability of other statements subsequent to conceptual change could then be calculated from the probabilities of members of the partition.

We have now considered four ways in which conceptual change can shift subjective probabilities thus influencing what we select as evidence. Conceptual change that shifts probabilities without altering the meaning of terms may enable a statement to compete favorably for selection that had failed to do so antecedently or may prevent a statement from competing favorably that had done so before. Such changes cannot affect which statements compete with which, but they can decide which statements are winners in the competition. The second kind of conceptual change alters the meaning of words and with it the relation of competition. Statements will either not compete with some statements they competed with before conceptual change or they will compete with some statements which they did not compete with before. Such conceptual change will also alter the probabilities. Hence the second kind of conceptual change, involving alterations in the meaning of words and the semantic status of statements containing them, should be expected to produce the most radical revisions in what statements are selected as evidence. These results of RE and RED coincide with presystematic intuitions. Fundamental conceptual revolution that changes the meaning of our basic scientific vocabulary is expected to bring with it a complete reconsideration of what should be counted as scientific evidence.

XVI. OBSERVATION AND THEORY

The foregoing remarks concerning the selection of evidence pertain equally to rule RE and RED. Conceptual change of the sort envisaged would influence what it would be reasonable to select as evidence according to either rule. However, there are some additional ways in which RED is sensitive to conceptual change. Theories interpreting scientific inference as inductive inference from observational evidence to laws and theories have been criticized by those seeking to emphasize the importance of conceptual change in scientific inference on the grounds that observational evidence is itself based on theoretical assumptions. Rule RED enables us to explicate the precise manner in which observational evidence is based on theory.

Suppose the doxastic system of a man contains the statement that he believes that he sees the path of an alpha particle in a cloud chamber. Now the statement that he *believes* that he observes this may be relevant to the statement that he does observe it. If he assigns high probability to a theory affirming that certain visible streaks of a clearly identifiable sort are the paths of alpha particles, then he may assign a high conditional probability to the statement that he is observing the path of an alpha particle relative to the statement that he believes that he is observing such a path. As a result of assigning a high probability to a theory, he estimates that the chances of his observing such a path when he thinks he does are very good. Moreover, if his subjective conditional probability of the observation statement on the belief statement is sufficiently high the observation statement may, as a result, be selected as evidence relative to his doxastic system containing the belief statement.

In general, by making the selection of evidence relative to the doxastic system of a man, RED enables us to explicate why observation statements are customarily selected as evidence and how the selection of those statements as evidence depends on theoretical assumptions. The doxastic system of a man will include his beliefs concerning what he observes. Most men will take their belief that they observe something as increasing the probability of the conclusion that they do observe it, but they will also consider some such beliefs to be more reliable than others.[22] When a man believes that he observes something at a great distance or under adverse conditions where he could easily mistake one thing for another, he will not consider such beliefs to be very reliable. Hence he will *not* estimate that the chances of such beliefs being true are especially high. On the other hand, when the conditions are ones he believes to be ideal for observing what he thinks he observes, then he will estimate that the chances of being correct about what he thinks he observes are very good. Whether a man estimates that the chance of his being correct in what he believes he observes are high will depend on what he believes to be ideal conditions for observing the kind of thing in question. His beliefs about what conditions are ideal for making observations of a given kind of thing constitute a hypothesis about what things of that kind are like. It is precisely laws and theories that supply such information. Therefore, when a man assigns so high a conditional probability to an observation statement relative to his belief in the truth of that statement that the observation statement is selected as evidence by RED relative to his doxastic system, it is because he assigns very high subjective probability to some general statement affirming that the conditions in which the belief arises are ideal for making observations of the sort in question.

Thus RED enables us to explicate why observation statements are so often prime candidates for evidence statements. It is because a man estimates that such statements have a greatly increased chance of being true relative to his believing them to be true. In this, observation statements contrast with general statements and theories. A man's belief that observation statements are true will often make them so highly probable on the basis of his doxastic system as to be selected as evidence by RED, while his belief that general statements are true may not affect the probability of those statements strongly enough to project them into evidence.

However, the foregoing remarks are not intended to suggest that general statements can not attain the status of evidence. Such a statement may attain this status, not because its probability is greatly increased relative to the statement that the man in question believes it to be true, but for some other reason. First, the man may estimate that the statement has a good chance of being true, whether or not anyone believes it, and thus assign it a very high antecedent, that is, unconditional probability. He may be influenced in this assignment by a variety of scientific and conceptual considerations. The high antecedent probability assigned to the statement may justify the selection of the statement by RED because it sustains a high conditional probability relative to his doxastic system.

The second reason for selecting a general statement as evidence relative to doxastic system concerns the influence of social consensus and authority. The doxastic system of a man includes beliefs about what others believe and beliefs about how reliable they are in various matters. For example, if a man believes that almost all reputable scientific specialists in a certain area of expertise believe a general statement to be a law, he may estimate that there is a good chance of the statement being true relative to his belief that these authorities consider it a law. The subjective probability he assigns to the general statement relative to his belief that the experts believe it to be a law may be so high that the general statement is selected as evidence by RED relative to his system.

XVII. INTERSUBJECTIVE RATIONALITY

The preceding argument illustrates some intersubjective influence on the selection of evidence by rule RED. This intersubjective influence is, however, exerted entirely through the subjective conviction of the man in question. We must now squarely face the question of the proper role of intersubjective restraint on the selection of evidence by our rules. We may allow that it is reasonable for a man to select evidence statements by RE and RED on the basis of his assignment of subjective probabilities, but we may wish to deny that his probability assignment is reasonable at the present stage of inquiry.

From our subjective starting point, it is a problem to explain in what sense the subjective probability assignment of a man can be unreasonable when it satisfies the calculus of probability. We cannot reject an assignment of subjective probabilities by appeal to objective probabilities. To do so would be to place the cartload of conclusions before the workhorse of evidence. The solution to the problem is to appeal to the subjective probabilities of others who search after truth as a restraint on the subjective probabilities of the individual. The rationality of the subjective probabilities of the individual would then be made to depend on agreement with the subjective probabilities of other truth seekers. But what sort of agreement is required? It would be implausible to require exact numerical agreement. We might require agreement within some interval, but the selection of end points of such an interval would be arbitrary. Instead, we shall require that a rational subjective probability assignment preserve the appropriate kind of *relevance* between statements when there is agreement among those who search for truth concerning such relevance.

The kind of relevance germane to our discussion may be defined in terms of positive and negative relevance. A statement h is positively relevant to k at time j if and only if $p_j(k, h)$ exceeds $p_j(k)$, and h is negatively relevant to k at time j if and only if $p_j(k)$ exceeds $p_j(k, h)$. A statement h is relevant to k at time j if and only if h is either positively or negatively relevant to k at time j.[23] With relevance thus defined, let us consider the probability assignments of those engaged in the search for truth at a given state of inquiry. They may disagree in their probability assignments and in their assignments of positive and negative relevance. However, consider the set of relations of positive and negative relevance, R_j, on which they do agree. This set, R_j, represents the area of intersubjective agreement concerning relevance at time j. Consequently, we may require of an intersubjectively rational assignment of probabilities that it preserve the relations of relevance in R_j. This requirement is embodied in the following condition:

> ISA. A probability function p_j is intersubjectively rational if and only if it satisfies the calculus of probability and the relations of relevance in R_j.

The requirement that a rational assignment of subjective probabilities satisfy the calculus of probability was justified earlier. How restrictive ISA turns out to be will vary from time to time, and will depend at a given time on how much intersubjective agreement there is concerning positive and negative relevance at that time.

Requirement ISA allows that a man may assign subjective probabilities in a way that yields many relations of relevance not prescribed by R_j. A man is not limited by ISA to the relations of relevance agreed upon by other investigators. This leaves some room for the cognitive iconoclast who discerns relations of relevance where others do not. This concession to iconoclasm is not a free ticket to speculation because even the iconoclast is restrained from repudiating relations of relevance agreed upon by other inquirers. What, then, of the man who has some theoretical or experimental insight which leads him to fly in the face of this restraint? On our account, his subjective probability assignment is not rational. This is not implausible. A man who denies relevance where all others who seek after truth affirm it is considered to be irrational. Of course, if his views prevail, then at some subsequent time it may become irrational by ISA for men to assign probabilities affirming relations of relevance which made his subjective probability assignment intersubjectively irrational in his own day. Such are the harrowing vicissitudes of inquiry.

In turning from RE to RED, it might seem reasonable to restrain the beliefs belonging to a doxastic system to those that agree in some way with the beliefs of others. As we noted earlier, subjective probabilities may be construed as beliefs about objective probabilities, in which case ISA would constitute a restraint on belief. But other restraints are unnecessary. Whether a man selects a statement as evidence by RED is not determined by what he believes but by the subjective conditional probability of the statement being true relative to his doxastic system. Having already placed an intersubjective restraint on the rational assignment of subjective probabilities, the influence of beliefs on the selection of evidence is already restrained by a condition of intersubjective agreement. The rational selection of evidence must strike a balance between the soboriety of intersubjective agreement and the inventiveness of subjective individuality.

XVIII. SUMMARY AND CONCLUSION

We have now completed our account of the selection of evidence under the influence of conceptual change. We argued that it is reasonable to select as evidence all those statements whose selection gives us positive value in the search for truth. Such value was defined in terms of empirical content. The empirical content of a statement was defined in terms of the subjective probability of its strongest competitor. The negative value assigned to selecting a false statement was equated with the subjective probability of the strongest competitor of that statement, while the positive value of selecting a true statement as evidence was equated with the subjective probability of the denial of the strongest competitor of that statement. Thus, the expected value of selecting a statement as evidence is determined entirely by the assignment of subjective probabilities. The only restraints we imposed on the assignment of subjective probabilities were those of logic, consistency and coherence, and that of intersubjective agreement concerning relevance.

Shifts in our assignment of subjective probabilities affect what it is reasonable to select as evidence. Such shifts may result from a variety of influences. Foremost among these is conceptual change. When conceptual change brings with it some alteration in the meaning of terms and the semantic status of statements containing them, then the subjective probabilities of all contingent statements of the language may be shifted thereby. Moreover, such alteration alters the competitive relations between statements as well as their subjective probabilities.

Other factors that may influence our assignment of subjective probabilities include the special features of theories, such as simplicity, completeness, and coherence; the stimulation of our senses, in the laboratory and everyday life; and the persuasion of our peers, both at our elbow and in the press. In our account, everything turns on subjective probabilities and beliefs. As subjective probabilities shift under the influence of these factors and others, our selection of evidence statements shifts as well. Some statements are added to evidence. Others are deleted. The shifting sands of subjectivity shape and reshape the foothill paths of evidence that guide us to conclusions in the mountainous terrain of inquiry.

University of Rochester

NOTES

* Research for this paper was supported by the National Science Foundation.
[1] See Sellars (1948).
[2] For a collection of articles by these authors on subjective probability, see Kyburg and Smokler (1964), and Jeffrey (1965).
[3] See Putnam (1967), p. 113.
[4] See Levi (1967), pp. 121-124.
[5] See Carnap (1962), pp. 168-175.
[6] See Reichenbach (1949), pp. 72-76.
[7] Cf. Levi (1967), pp. 56-74.
[8] See Hempel (1962), pp. 98-169, Hintikka and Pietarinen (1966), pp. 96-112, Hilpinen (1968) and Levi (1967).
[9] *Ibid.*
[10] For a discussion of such conceptions, see Hintikka (1968), pp. 311-332.
[11] See Feyerabend (1962), p. 50.
[12] See Carnap (1962) pp. 70-72. Carnap, however, concerned with logical rather than subjective probabilities, only proposed the identification of certain of the former with estimates in special cases.
[13] See Hintikka (1964), pp. 274-288.
[14] I defend this thesis in Lehrer (forthcoming).
[15] I use the terms 'contradiction' and 'logical truth' as Carnap used 'L-false' and 'L-true' so that a statement is a contradiction if the falsity of the statement follows from the rules of logic and the meaning postulates and is a logical truth if the truth of the statement follows from such principles. See Carnap (1952), pp. 65-73.
[16] In earlier papers the rule was formulated in such a way as to include logical truths in the evidence, but the exclusion of logical truths from the evidence does not alter the results cited. See 'Induction and Conceptual Change' (Lehrer 1970a) for a formulation of the rule and the appendix of Lehrer (1970b), pp. 127-131, for the relevant proofs with respect to a similar rule.
[17] Cf. Lehrer (1970b), pp. 130-131.
[18] The term 'doxastic' is taken from Hintikka (1962), p. 48.
[19] See Levi (1967), p. 59.
[20] This is commonly assumed by defenders of subjective probability (see, for example, the articles Kyburg and Smokler (1964)), but not by Jeffrey (see his (1965) pp. 153-170).
[21] See Jeffrey (1965) pp. 153-170, esp. p. 158. For a discussion of Jeffrey's procedure see Levi (1967-68), pp. 197-209, Harper and Kyburg (1970) and Jeffrey's reply in (1970).

However, it is clear from the latter article that Jeffrey rejects the idea of selecting evidence statements.

[22] The relevance of a perceptual belief to the evidential status of the statement believed was, to my knowledge, originally proposed by Chisholm (1957), p. 82. But his approach is less subjectivistic. See p. 112.

[23] Cf. Carnap (1962), p. 348.

BIBLIOGRAPHY

Carnap, R., 1952, 'Meaning Postulates', *Philosophical Studies* **3**, 65–73.
Carnap, R., 1962, *The Logical Foundations of Probability*, Chicago.
Chisholm, R. M., 1957, *Perceiving: A Philosophical Study*, Ithaca.
Feyerabend, P., 1962, 'Explanation, Reduction, and Empiricism', in *Minnesota Studies in the Philosophy of Science*, Vol. III (ed. by H. Feigl and Grover Maxwell), Minneapolis, pp. 28–97.
Harper, W. and Kyburg, H. E., 1968, 'Discussion: The Jones Case', *British Journal for the Philosophy of Science* **19**, 247–51.
Hempel, C. G., 1962, 'Deductive-Nomological vs Statistical Explanation', in *Minnesota Studies in the Philosophy of Science*, Vol. III (ed. by H. Feigl and G. Maxwell), Minneapolis, pp. 98–169.
Hilpinen, R., 1968, 'Rules of Acceptance and Inductive Logic', *Acta Philosophica Fennica* **22**, Amsterdam.
Hintikka, J., 1962, *Knowledge and Belief*, Ithaca.
Hintikka, J., 1964, 'Towards a Theory of Inductive Generalization', in *Logic, Methodology, and Philosophy of Science*, Vol. II (ed. by Y. Bar-Hillel), Amsterdam, pp. 274–288.
Hintikka, J., 1968, 'The Varieties of Information and Scientific Explanation', in *Logic, Methodology, and Philosophy of Science*, Vol. III (ed. by B. van Rootselaar and J. F. Stall), Amsterdam, pp. 331–332.
Hintikka, J. and Pietarinen, J., 1966, 'Semantic Information and Inductive Logic', in *Aspects of Inductive Logic* (ed. by J. Hintikka and P. Suppes), Amsterdam, pp. 96–112.
Jeffrey, R., 1965, *The Logic of Decision*, McGraw-Hill, New York.
Jeffrey, R., 1970, 'Acceptance vs Partial Belief' in *Induction, Acceptance, and Rational Belief* (ed. by Marshall Swain), D. Reidel, Dordrecht, pp. 157–85.
Kyburg, H. and Smokler, H. (eds.), 1964, *Studies in Subjective Probability*, New York.
Lehrer, K., 1970, 'Justification, Explanation, and Induction', in *Induction, Acceptance, and Rational Belief* (1970) (ed. by Marshall Swain), D. Reidel, Dordrecht, pp. 100–33.
Lehrer, K., 1971, 'Induction and Conceptual Change', *Synthese* **23**, 206–25.
Lehrer, K. (forthcoming), 'Induction, Rational Acceptance, and Minimally Inconsistent Sets', in *Minnesota Studies in the Philosophy of Science*.
Levi, I., 1967, *Gambling with Truth*, New York.
Levi, I., 1967–68, 'Probability Kinematics', *British Journal for the Philosophy of Science* **18**, 197–209.
Putnam, H., 1967, 'Probability and Configuration', in *Philosophy of Science Today* (ed. by Sidney Morgenbesser), New York, pp. 100–14.
Reichenbach, H., 1949, *The Theory of Probability*, Berkeley and Los Angeles.
Sellars, W., 1948, 'Concepts as Involving Laws and Inconceivable Without Them' *Philosophy of Science* **15**, 287–315.

STEPHAN KÖRNER

LOGIC AND CONCEPTUAL CHANGE

The systems in whose changes we are interested are systems of thought. I shall devote the first part of this essay to a general description of their structure; to a distinction between its various aspects and to the manner in which they may suffer or resist modification; and lastly to a brief consideration of the causes of conceptual change. The remarks made in the first part of this essay will then in its second part be exemplified by, and applied to, logic.[1]

I

For our present purpose it will be convenient to collect under one name some of the more general, more common and apparently more stable aspects of systems of thought; and to contrast them with their more specific, less widely shared and apparently less stable features. The contrast may be expressed as a distinction between the logico-categorial, or, briefly, categorial framework of a system of thought and its content. But it must be borne in mind that the distinction between framework and content might be drawn in different ways, depending on the purpose served by the distinction. The notion of a categorial framework can be explained both adequately and briefly if we adapt some traditional logico-philosophical notions. The adaptation consists in each case in replacing an absolute notion by a relativized version of it.

We accept the distinction between particulars and attributes, without however assuming that everybody draws it in the same manner. Thus both Plato's numbers, which are particulars, and Russell's numbers, which are not, are regarded as legitimate conceptions. We accept the (Aristotelian) notion of maximal kinds or *summa genera*, without however assuming that of all the possible classifications of entities into natural kinds, only one can be correct. We accept the (Kantian) notions of constitutive principles in the sense of propositions to the effect that being a member of a certain maximal kind logically implies possessing a certain attribute; and of individuating principles in the sense of propositions to the effect that being a distinct member of a certain maximal kind logically implies possessing a determinate subattribute of a certain more general attribute. But we do not assume that of the possible sets of constitutive and individuating principles, associated with a given maximal kind, only one set is the correct one. Thus we admit for the maximal kind, if any, of external phenomena either a principle of causal, or, for example, a principle of probabilistic connection, as constitutive; and a principle of location in either Euclidean space and time, or, for example, of location in a four-dimensional space-time continuum, as individuating. We accept, finally, the (Aristotelian) distinction between independent particulars, which exist independently of the existence of other particulars, and dependent particulars which exist only as features of independent particulars. But we do not assume that there is only one correct way of making the distinction and regard as legitimate both Aristotle's conception of a horse as an independent particular or, in his terminology, as a primary substance, and Spinoza's concept of it as a dependent particular.

It should be noticed that in employing the notions and distinctions which have just been explained, one *eo ipso* employs a concept of logical implication and thereby, a logic by the rules of which valid implications and quite generally valid propositions are distinguished from invalid ones. That a particular belongs to one maximal kind of an accepted natural classification logically implies something about its membership of the other maximal kinds. Again the notion – or, more precisely, a notion – of logical implication, and with it a logic, enters the assumption of any constitutive or individuating principle. Lastly, of course, the meaning of any concept depends on what is logically implied by its applicability to an object.

A suitable notion of a categorial framework can now be explained. To employ a categorial framework is to employ (i) a distinction of the features of experience into particulars and attributes; (ii) a categorization of the particulars into maximal kinds of independent particulars and of dependent particulars, if any; (iii) a set of constitutive and individuating principles associated with each maximal kind; (iv) an underlying logic. In proposing this explicit definition of the notion of a categorial framework, I do not wish to imply that everybody is at any time fully aware of his categorial framework, that he may not be confused or wholly undecided about some of its aspects, or that he may not waver between two or more different frameworks. Such conclusions do not follow from our definition. They are, moreover, incompatible with the history of ideas and with the history of many people's intellectual development.

Two systems of thought may have the same categorial framework, and yet differ in a great variety of ways. The differences may in particular concern their finer classificatory structure, i.e. the subclassification of their maximal kinds into species and subspecies of which some may be associated with more specific constitutive and individuating principles. Two systems with the same categorial and even with the same classificatory structure may still greatly differ in the beliefs accommodated by these structures, especially in the non-logical relations (correspondence and partial correlations) which are assumed to hold between the instances of their maximal kinds and species.

As regards the possible changes of a system of thought we may distinguish modifications which do not affect its categorial or even its more specific classificatory structure – for example a change of belief about the presence or absence of a contingent relation between every member of one class and some member of another class of particulars; modifications which affect the classificatory structure of the system of thought without affecting its categorial structure – for example a taxonomical revision of the species of an animal genus; modifications which affect the categorization, the constitutive or individuating principles of a categorial framework without affecting its underlying logic; and lastly modifications affecting this logic. That people employ the distinctions and notions which have been collected under the name of a categorial framework is an empirical fact. And just as our most remote animal ancestors apprehended the world without employing a categorial framework, so our post-human descendents, if any, might again apprehend it quite differently.

Categorial frameworks may be, and have been, modified. The modification may amount to a straightforward replacement of a concept, set of concepts, or other feature of a categorial framework by some alternative, as when for example an ontology admitting mental and physical substances is replaced by an ontology admitting only one kind of substance. The modification may, however, be not so much a replacement of one feature by another, as an adjustment between an old feature and a new one, which to some extent allows the retention of the former, as well as the introduction of the latter. Such adjustments account for the historical fact that conceptual change may be gradual. In distinguishing between various kinds of conceptual change in which new features are adjusted to old ones, I neither wish to exclude the possibility of hybrid varieties nor to claim completeness.

A fairly simple way of achieving innovation without replacement and of avoiding conflict between an old feature and a new one, is to construe one of them as ontologically fundamental or primary and the other as auxiliary or secondary. As an example we may consider on the one hand an ontology implicit in a society's commonsense, as it is shared by laymen and experts and expressed in a natural language; on the other hand the esoteric ontology of some theory which is accessible only to a few experts and expressed in a technical extension of their natural language. Of the two ontologies each may be regarded as primary, and the other as secondary. People whose reflection on this conflict is influenced by the British tradition of empiricist, commonsense and linguistic philosophy, will regard the commonsense ontology as primary; while people who by instinct or education tend towards rationalist reconstruction will regard commonsense as temporally prior, but ontologically secondary, to one or more scientific theories or to some comprehensive metaphysical system.

A more subtle and complex way of achieving innovation without replacement is to incorporate the old features as parts or aspects into the new features or *vice versa*. Thus a person who undergoes a conversion from an Aristotelian to a Cartesian ontology may now conceive a horse, which he formerly regarded as a primary substance, as merely as aspect of Descartes' material substance. A person who ceases to believe in spirits as a maximal kind may construe them as aspects of persons who believe in them. More generally, my independent particulars exist in the world; those of your independent particulars which I do not acknowledge, exist merely as aspects of you – whether or not you yourself are an independent or a dependent particular. The demotion of old independent entities to mere aspects of new independent entities shows itself in a reclassification of the corresponding complete or categorematic symbols into incomplete or syncategorematic ones.

The desire to adjust, in the manner just described, an old ontology to a new one arises at times of radical conceptual innovations in conservative minds who have the ability to understand both ontologies and strong motives not to abandon either. An example is Leibniz who felt the conflict between the "substantial forms" and "mechanism" at the age of fifteen and decided first for mechanism and later for the substantial forms. In a famous letter written in 1714 to M. Remond de Montmort he not only describes the conflict but also indicates the reconciliation through the alleged discovery "*que les monades ou les substances simples, sont les seules véritables substances; et que les choses matérielles ne sont que des phenoménes, mais bien fondés et bien liés*".[2] A similar adjustment between the ontology of quantum theory and classical physics in which the phenomena described by quantum physics are "well founded" in a deterministic ontology is outlined by Einstein who confidently assumes that "the statistical quantum theory" will, "within the framework of future physics, take an approximately analogous position to the statistical mechanics within the framework of classical mechanics".[3]

Finally, conflicts between a traditional and a novel feature of a categorial framework can be resolved without ontological subordination of one to the other. In order to illustrate this possibility, it is useful to define generalizations of the concepts of (logical) incompatibility and of identity. Let us call two propositions "conflicting" or "incommensurable" if, and only if, the assumption that they describe the same entity, set of entities or possible world is internally inconsistent. Let us call two propositions "identifiable" (in certain specified contexts) if, and only if, they can (within the specified contexts) be treated as if they were identical. In accordance with these definitions logical incompatibility is a special case of incommensurability and identity a special case of identifiability. Putting it more explicitly, if two propositions are incompatible, they are *a fortiori* incommensurable; and if two propositions are identical (strictly speaking, if the propositional content of two expressions is identical) they are *a fortiori* identifiable (the propositional content of either expression is identifiable with that of the other expression). On the other hand, incommensurability does not imply incompatibility, nor does identifiability imply identity. Thus two propositions of which one is a commonsense description of a falling stone, while the other is a corresponding theoretical state-description of, say, the world of classical physics, are incommensurable. They are also in certain contexts – e.g. when we wish to estimate the time it takes for the stone to fall a given distance – identifiable. Indeed such identifications (in the contexts for which classical physics is employed) constitute the application of classical physics to experience.[4]

If we assume that the two incommensurable propositions are true descriptions of the one and only real world, we are committed to the thesis of the "identity of incompatibles" or of the "coincidence of opposites" – a thesis which has never lost its attraction for one wing of metaphysics from Parmenides to Hegel and beyond. If we assume that of the two incommensurable propositions at most one is true of reality, we may in the manner indicated above, construe the other as secondary, auxiliary, as of merely heuristic value, i.e. as ontologically subordinate to the first. But we may also consider the incommensurable propositions as describing different possible worlds and, hence, as compatible; and as being only in certain contexts identifiable and hence as not identical. The mysterious or, at least, mystifying principle of the "identity of opposites" is then replaced by a rather pedestrian principle of the "identifiability of incommensurables" neither of which is ontologically superior to the other.

Between the paradox of the identity of incompatibles, which seems to stem from some mystical urge, and the almost trivial obviousness of the limited identifiability of incommensurables, which results from an analysis of the application of different theories to commonsense experience, lies the metaphysical position of "perspectivism". According to this position, it is impossible for any member of a pair of incommensurable propositions to be true of reality, because reality cannot be described by propositions. A system of propositions every one of which is incommensurable and yet also identifiable with at least one other, consists, on this view, at best of different perspectives of the one reality.[5]

What has been said, however sketchily, about the structure of categorial frameworks and their flexibility in combining tradition and novelty, throws some light on the causes of conceptual change. It should in particular emphasize the danger of overestimating the revolutionary force of experimental falsification or of philosophical argument. As regards the former, it must first of all be noted that what is falsified by experiments is not a categorial framework, but rather a scientific theory or system of empirical beliefs which presupposes the distinctions, categorizations and assumptions which together make up the framework of the theory or system of beliefs. Moreover, it is agreed even by orthodox falsificationists that if a theory is falsified by experiments, one may save its "core" by suitable *ad hoc* hypotheses. Yet, what is, and what is not to count as belonging to the core or as being *ad hoc* may change from one group of people to another. Adding epicycles to a geocentric physics in order to harmonize it with observations may appear as a blatantly *ad hoc* procedure to some people, and as the only truly rational procedure to others.

Philosophical arguments of the so-called analytical kind either exhibit the structure of the categorial framework under investigation, or else suggest with various degrees of explicitness its more or less radical reconstruction. They provide neither premises by which the reconstruction is logically implied nor conditions by which it is causally necessitated. The same holds for the bolder proposals of speculative metaphysics which aim not at local reconstructions but at global revolution.[6] The need to make an implicit, accepted categorial framework explicit and to compare it with one or more as yet unaccepted competitors, is greater in periods of intellectual conflict than in times of gradual change. This is why in times of such conflict all theoreticians, and not only philosophers, tend to be philosophical.

Although neither philosophical analysis and speculation nor experimental falsification constitute logically or causally necessary conditions of categorial change, they are not therefore wholly ineffective in bringing it about or in preventing it. They are, it seems to me, best understood as being *among* the "pressures" towards conceptual change or stability. The interaction and relative strength of these pressures reveal themselves to the backward looking historian and present a challenge to the social scientist. It is not the task of philosophy to measure the force of the pressures exerted by experimental tests or by philosophical argument. But it is its business to exhibit the conceptual relations between the beliefs which exert these pressures and the categorial frameworks on which they are exerted.

II

In considering some of the ways in which the logical structure of a categorial framework may change or remain stable under various pressures, it is well to have examples of logical systems in mind. The best known systems are, of course, elementary classical logic, i.e. propositional calculus, quantification theory and theory of equality, as formalized by Frege and his successors; and elementary intuitionist logic, as formalized by Heyting and others. Each of these systems, which I shall briefly call L

and *I* may be extended by admitting that an object may not only be a positive instance of a predicate (when it is correct to apply and incorrect to refuse the predicate to the object) or a negative instance of a predicate (when it is correct to refuse and incorrect to apply the predicate to the object), but also a neutral instance of a predicate (when it is equally correct either to apply or to refuse the predicate to the object). The extensions of *L* and *I* at which one thus arrives will respectively be called *L** and *I**.[7]

These four logical systems will, even without any detailed analysis, be sufficient to show that the kinds of radical or gradual modifications discussed in the preceding section may, and do, affect even the logical structure of a categorial framework. Let us consider first the possibilities and then some of the pressures working towards realizing them. It is not difficult to imagine the case of a straightforward replacement of one logical structure by another with all the consequential changes in the meaning of 'logically implies' in the relevant classificatory statements, constitutive and individuating principles. An example would be the replacement of *L* by *I* or *I* by *L*.

The case where an old logical structure is reconciled with a new one by regarding one of them as ontologically primary also presents no difficulty to the imagination. A person using *L** in his commonsense thinking who wishes to employ a theory embedded in *L* or in *I* might well regard the whole theory and thus *a fortiori* its underlying logic as a merely heuristic device, the use of which detracts in no way from the ontologically fundamental character of *L**. Examples of the method of achieving innovation without replacement by regarding the old logic as capable of interpretation in the new are perhaps less obvious. However, Gödel has shown how the classical logic can be interpreted by the intuitionist and Beth and others have shown how the intuitionist logic can be interpreted by the classical.[8] Lastly, what I have called the "principle of the identifiability of incommensurables" is also applicable to two logical structures – for example when one of two incommensurable scientific theories acknowledges the actual infinite of classical, the other the merely potential infinite of intuitionist, mathematics.

Let us now turn to examples of scientific and philosophical pressures towards changes in the logical structure of categorial frameworks. We are here more intertesed in the way in which scientific and philosophical reflection leads to the emergence of new logical structures as thought possibilities, than in the manner in which the authority of acknowledged experts leads to the acceptance of some of these structures and the rejection of others. Our main examples of scientific pressures will be quantum mechanics and taxonomy, our main philosophical example will be an analysis of constructive thinking. It should be noted that the distinction between scientific and philosophic pressures is not meant to be sharp or absolute. Indeed, since the exhibition, reconstruction and new construction of categorial frameworks are traditionally regarded as philosophical tasks, a scientist's reflection on possible alternatives to an accepted categorial framework is *ipso facto* also philosophical.

Quantum mechanics as conceived by Heisenberg[9] does *not* conflict with the classical logic L for the simple reason that L is its underlying logic. It does not even conflict with every version of the principle of causality which has appeared in the dominant tradition of Western thought, but only with a particular version of it. In order to make this clear, I shall distinguish between the general principle of causality and the (more precisely, *a* particular) epistemic principle of causality. The former is the thesis that in virtue of the (known or unknown) laws of nature the state of the universe at any time is completely determined by its preceding states and completely determines its succeeding states. The latter is, in Heisenberg's words, the thesis that "if we know the present exactly, we can calculate the future" (*loc. cit.* p. 197). He holds (1) that the epistemic principle means the same as the general principle of which it is merely "the exact formulation"; and (2) that quantum-mechanics is incompatible with the epistemic principle of causality and thus – provided that (1) is correct – with the general principle of causality. Strictly speaking, quantum mechanics is incompatible only with the antecedent of the epistemic principle and, hence, with the assumption of its applicability. For if quantum mechanics is true, then "we cannot as a matter of principle come to know the present in all its component details" (*loc. cit.*).

Heisenberg's reasons for equating the epistemic and the general principle of causality are based on a positivistic *credo*, which even among physicists is sectarian rather than ecumenical. If, unlike Heisenberg, one assumes that "behind the perceived statistical world there lies hidden a 'real' world in which the [general] principle of causality is valid", one may incorporate quantum mechanics into a categorial framework in which the class of events form a maximal kind with which the general principle of causality is associated as a constitutive principle. For one could then without inconsistency assume that, although every event is causally determined by one or more of its predecessors, it is humanly impossible to possess all the knowledge which is required for certain causal predictions or certain classes of them. We could then regard quantum mechanics as merely another example of one theory replacing another within the same categorial framework.

On the other hand, because quantum mechanics, unlike classical physics, is incompatible with the antecedent of the epistemic principle of causality, it presents, or at least clearly suggests, thought possibilities whose realization radically modifies the earlier categorial structure. The modifications which are suggested by the assumption that there are unavoidable gaps in men's knowledge of causal connections are, first, that nature itself contains such gaps, and, secondly, that the law of excluded middle, and thus L, is not valid. The road which leads from epistemic to material and from material to logical indeterminacy has been actually taken by some philosophers and physicists. The first step has in fact been taken by Heisenberg himself. For whatever we may think of the positivist and, for that matter the Kantian, absorption of ontology into epistemology, quantum mechanics fits well into a framework in which the general principle of causality is not associated with the class of events as one of its constitutive principles. The assertion that quantum mechanics is possible only within a causal framework is just as mistaken as the assertion that it is possible only within a non-causal one. One can without abandoning classical logic assume the absence of material (as opposed to merely epistemic) causal connections in certain regions or strata of the universe and assume that these regions or strata are subject to objective probabilistic laws.

Material indeterminacy, which results from abandoning the principle of general causality, implies epistemic indeterminacy, which results from abandoning the principle of epistemic causality. But epistemic indeterminacy does not imply material indeterminacy. Again, logical indeterminacy, which results from abandoning the law of the excluded middle, implies material indeterminacy without being implied by it. By logical determinacy we understand the general validity of the law of excluded middle or, more specifically, its validity for temporal propositions of the form: 'At time t such and such a certain determinate region of the universe has such and such a determinate characteristic.' According to the law of excluded middle, this statement is either true of false (for all time) independently of whether the temporal reference *in* the statement refers to the past, present or the future. To assume logical indeterminacy for temporal statements is to abandon the law of excluded middle and to assume instead that some statements containing a reference to an as yet unrealized future state of the universe are as yet neither true nor false. Such an assumption has been made, e.g., by Reichenbach.[10] The logic by which L is replaced in discussions of quantum mechanics is usually a three-valued logic, in which the *tertium non datur* is replaced by a *quartum non datur*. But a suggestion of replacing L by I appears no less reasonable.

My next example of a pressure towards replacing an accepted logical structure by a new one, or at least, towards logical innovation without replacement, comes from taxonomy. In the system of Linnaeus, species are regarded as, among other things, constant and clearly separable, even though Linnaeus eventually inclined to the view that some species might have arisen since the creation as a result of hybridization.[11] Let us call a hybrid a "common border-line case" of its respective "parent-species" if, and only if, the parent species are mutually exclusive and the hybrid can with equal correctness be judged to be a member of one or the other of the parent-species. The statement that such objects exist is from the point of view of L (and for that matter of I and any other "exact" logic) logically impossible. It is, however, logically possible from the point of view of L^* which admits such objects in principle. The desire to account for cases of continuous "intergrading" between some species constitutes a pressure towards the explicit or implicit acceptance of L^* as one's primary or, at least, a secondary logic. But there are other ways of accounting for apparent border-line cases of two parent-species – for example, by postulating that all individuals of the two species and their apparent common border-lines case are located on a partly actual and partly potential continuum of the Dedekindian type.

If for some reason one regards commonsense as more fundamental than science, then the taxonomic pressure towards adopting L^* as one's primary logic might be reinforced by an analysis of the logical structure of commonsense thinking. Such an analysis would show that commonsense thinking allows for border-line cases not only between different kinds of animals, but also between many other kinds of objects. (It is worth noting here that the commonsense of different societies and different epochs is rich in suggestions of logical reconstruction. One of them derives from an analysis or our practical commonsense thinking.)

Practical or constructive thinking is thinking about the way in which one of a number of mutually incompatible states of affairs – all of them apparently realizable and none of them as yet realized – can be brought into existence. An analysis of constructive mathematics, as conceived and practised by the intuitionist mathematicians, shows that its underlying (primary or secondary) logic is I. Constructive mathematical thinking can be regarded as a special case of constructive thinking, which ignores border-line cases. Again, the logic of constructive, commonsense thinking, as expressed in a natural language, is I^* or at least very similar to it. Since the logic underlying other regions of thinking, such as classical mathematics and physics, is L and since a great deal of commonsense thinking has L^* as its underlying logic, the question arises whether I (or I^*) and not L (or L^*) is, could be, or should be our primary logic. Until the explicit formulation of intuitionist logic by intuitionist mathematicians, L used to be considered quite generally as the primary logic or at least a substantial part of it.

Even if the mere confrontation with an explicitly formulated system, I, could not be regarded as a pressure towards accepting it as one's primary logic, it at least clears the path for other pressures in this direction. A person who already believes that all thinking is somehow constructive will, if he comes to accept the analysis of constructive thinking as subject to I (or I^*), be strongly inclined to accept this system as his primary logic. The same will be true of a person who already believes in Kantian fashion that theoretical reason is subordinate to practical reason and whose reflections on I (or I^*) make him regard it as the logic of practical reasoning. Again just as quantum mechanics suggests a transition from epistemic to material and from material to logical indeterminacy, so can reflection on the nature of choice, especially choice under prudential or moral constraints, suggest a similar transition which may lead to the rejection of the principle of excluded middle and the acceptance either of I or I^* or some other logical system in which the principle is not valid. A still different kind of supporting pressure may come from a mathematics – conceived not as an idealization of the actual world but as a description of its structure – in which the rejection of actually infinite sets is associated with the rejection of non-constructive proofs based on the principle of excluded middle.

All such pressures are resistible. The methods which are available in implementing radical change, gradual modification or total preservation of the non-logical aspects of a categorial framework exposed to intellectual pressures are, as we have seen, also available when the pressures are directed towards its logical structure. Moreover, even a much better understanding of the structure of categorial frameworks, and of the methods of implementing or resisting their change under intellectual pressures of various kinds, would not imply an understanding of the laws, *if any*, which determine when the pressures are, and when they are not, resisted.

It seems proper to conclude these remarks by repeating a warning against the frequent mistake of confusing the implementation of a prior ontological thesis concerning the maximal kinds, constitutive and individuating principles, and the logical structure of an accepted or proposed categorial framework on the one hand, with a demonstration of the thesis on the other. The warning is not out of place, since it would apply even to Brentano and Russell who, by their theories of complete and incomplete (categorematic and syncategorematic) symbols, have greatly increased our understanding of some of the implements by which new ontological convictions are adjusted to old ones. They both put, as it were, the cart containing their methodological implements before the horse of their ontological convictions, namely a "reist" ontology embedded in L in Brentano's case, and a sense-data metaphysics also embedded in L in the case of Russell.

University of Bristol

NOTES

[1] For a more detailed explanation of the notions introduced in Part I, see Körner (1970) Part I.
[2] Gerhardt (1875), v. 3, p. 606.
[3] Schilpp (1951), p. 672.
[4] See Körner (1966), esp. Part III.
[5] For more details see Körner (1971).
[6] Körner (1969).
[7] For details of the structures of L^* and I^* and their relations to L and I, see Cleave (1970).

[8] For some of the relevant literature, see e.g. Rasiowa and Sikorski (1968).
[9] Heisenberg (1927).
[10] Reichenbach, (1948).
[11] See, e.g., Cain (1959).

BIBLIOGRAPHY

Cain, A. J., 1959, 'The Post-Linnaean Development of Taxonomy' in *Proceedings of the Linnaean Society of London*, Vol. 170, 1970th session, 1957–58; Part 3, April 1959, pp. 234–44.
Cleave, J. P., 1970, 'The Notion of Validity in Logical Systems with Inexact Predicates'' *British Journal for the Philosophy of Science* 21, 269–74.
Gerhardt, C. I., 1875, *Die Philosophischen Schriften von Gottfried Wilhelm Leibniz*, Berlin, pp. 1875-90.
Heisenberg, W., 1927, Über den anschaulichen Inhalt der quantentheoretischen Kinematik und Mechanik', *Zeitschrift für Physik* 43, 172–98.
Körner, S., 1966, *Experience and Theory*, London.
Körner, S., 1969, 'Categorial Change and Philosophical Argument' in *Proceedings of the Israel Academy of Sciences and Humanities* 3, 255–69.
Körner, S., 1970, *Categorial Frameworks*, Oxford.
Körner, S., 1971, *Abstraction in Science and Morals* (Eddington Memorial Lecture), Cambridge.
Rasiowa, H. and Sikorski, R., 1968, *The Mathematics of Metamathematics*, 2nd ed., Warsaw.
Reichenbach, H., 1948, *Philosophic Foundations of Quantum Mechanics*, Berkeley and Los Angeles.
Schilpp, P. A. (ed.), 1951, *Albert Einstein: Philosopher-Scientist*, New York.

JOSEPH S. ULLIAN

SOME COMMENTS ON PROFESSOR KÖRNER'S PAPER

With sweeping strokes Professor Körner has painted us some metaphilosophy. Portions look insightful and suggestive, if viewed from favourable angles; other portions, to which my comments are directed, betray an overthick brush and invite illusion. The shape of the whole is that of the outline of a gridwork, a Procrustean crosshatch on which to lay belief bodies.

Professor Körner's categorial framewoks are cast in terms that are rather too Aristotelian for my own taste, but I am not greatly bothered by that. My first really bad moment comes when he says (p. 124) that

the notion - or, more precisely, a notion - of logical implication, and with it a logic, enters the assumption of any constitutive or individuating principle.

For this suggests that we are to think of logic*s* – not just different formal theories, but different logics. This augurs ill.

And so does the remark (p. 124) "I do not wish to imply that everybody is at any time fully aware of his categorial framework...". One gathers that, for Körner, a categorial framework is itself like a system of beliefs, and that there can be objective criteria for determining just what categorial framework a given individual, or school, has. Further, he thinks that "a scientific theory or system of empirical beliefs" may presuppose "the distinctions, categorizations and assumptions which together make up the framework of the theory or system of beliefs" (p. 128). And this seems to say that such a theory or system may presuppose a particular framework. That is unclear to me. What I could make much better sense of would be the claim that a theory or system may be more comfortably characterizable – or perhaps even articulable – in one philosophical vernacular than in another. Ascriptions of frameworks to theories, if I am to understand them at all, had best be taken as rather grand ways of talking about comfort of characterization, or goodness of fit. For me, then, that a theory rests on a particular framework comes to saying that that framework's distinctions let us give a good account of the theory.

Pearce and Maynard (eds.), Conceptual Change, 137–140. *All rights reserved.*
Copyright © 1973 by D. Reidel Publishing Company, Dordrecht-Holland.

And since anything that can be said can be said in lots of ways, nothing even approaching uniqueness of framework is ever to be hoped for. I see this as clarification (and maybe de-metaphysicalization) by semantic ascent.

Had its "underlying logic" not been counted part of a categorial framework the outright attribution of frameworks to theories might have troubled me less than it did; yet still some, given the cast of the rest of the cast. But the notion that different belief systems may actually embody different logics is very difficult for me to make sense of. I grant that formal systems may differ, terms may differ, and techniques of argument and their upshots may differ. Still, when looking for "a logic" in the philosophical bush I import my own. Someone else's method of reasoning, or of making distinctions, must be seen in the light of my own of it is to be seen by me at all.

In the same vein: "replacement of one logical structure by another" (p. 130) might happily give way to "replacement of one formal logical theory by another"; "the consequential changes in the meaning of 'logically implies'" (p. 130) then becomes " the consequential changes in the formal relation of deducibility." While I do not know what "new logical structures as thought possibilities" (p. 130) might come to, I at least can envisage new species of formal apparatus. So on one level, if not on another, I can trace paths parallel to some of Körner's strokes.

Only by ascent to the level of discourse *about* logic can I make sense of such a suggestion as that the law of excluded middle might not be valid. I can imagine codifications of deductive practice that might differ extensively from the codifications common nowayears; I can imagine codifications, or articulations of theories, in which something might be labeled "the law of excluded middle" and then exorcised. More generally, I can imagine articulations that might be very comfortably embedded in broader contexts where something identified as that law was denied. In terms such as these I could no doubt succeed in finding a construal for the suggestion that the law be relinquished. But from *my* frame of reference, even the most radical of articulations will be understood, if at all, by way of logical terms and concepts familiar to me. Discourse about logic is, after all, still discourse.

In the open discussion at this conference session Hilary Putnam chided me for wanting to maintain "a thread of unrepresented absolutism", if I remember his phrase correctly. I suppose the charge is that I want to take L as *being the logic* regardless of what body of beliefs might seem acceptable and regardless of what articulations for those beliefs might seem appropriate. This is partly right and partly wrong. I certainly don't want to say that there are any claims that are well seen as totally immune to revision, any claims that are absolutely privileged. This applies to claims that might be regarded as belonging to logic, so it applies to enunciations of principles of L. But I also want to say something more positive. I want to insist that I cannot conceive of there being a theory or belief system intelligible to me for which I could not comfortably construct a *metalanguage* whose explicit logic was L. I say "explicit logic" because I still do not find very good sense in talk of what logic a theory "has", unless something labeled "logic of the theory" is given with it. I am not claiming that alternative explicit logics could not be used, only that L could be. Finally, I am not even holding that the claims I am making here should be seen as totally immune to revision, embrace them enthusiastically though I do. If all this leaves me an Absolutist then I am just stuck with that.

Back to the "logics" of categorial frameworks. I would construe acceptance of L^* as adoption of a style of discourse, formalized or not, that was seen to lend itself to ready characterization in L^*'s terms – that was smoothly codifiable in a metalanguage giving lip service to L^*. Nor could I imagine what else to make of it. Talk of *pressure* toward the acceptance of L^* seems to me misleading. I can only conceive it on analogy with "pressure" to shift from rectangular to polar coordinates for certain technical purposes. In a recent paper[1] David Sanford studies the grue-bleen problem in an exciting and novel way, giving extensive consideration to borderline cases of predicates. Unsurprisingly, his work can be nicely accommodated without even apparent departure from L. It can also be accommodated within a somewhat different formulation of "logic", perhaps L^*. Which is the *underlying logic* of his treatment? The question surely yields, here and elsewhere, to the question of how *we* want to characterize his apparatus. In the least misleading sense of the words, the *underlying logic* of any theory, as understood by us, is that of *our* conceptual scheme. What can and does vary with theory might better be seen as *surface logic*, revealed mainly by mode of articulation.

Novel formulations, novel ways of talking, can serve us greatly.

But this should be no solace for those who would have us think that even the core of our conceptual scheme is ours to shed at will, or to shed at all. As Quine says, "There is no cosmic exile", no one to stand outside his conceptual scheme and assess it unencumbered by a way of thinking.

I must say that I find no temptation to agree with Körner when he says that constructive mathematics can be taken as subsumable under constructive thinking. Parlaying these two uses of "constructive" into an entry seems like mixing a greyhound with a chow chow. Only such soft ground as this can I find for the mix (since I am no fan of "open futures"): We can imagine a case in which it would help us not at all to know that one among us was a spy unless we knew which one he was. So we would accept no "existence proof" that was nonconstructive. But even then, we could surely imagine ourselves to have obtained the former useless piece of information while still remaining ignorant of the spy's identity. There's a spy; perhaps it's Ortcutt, perhaps it isn't.

That I is the underlying logic for constructive mathematics speaks to me only of the smoothest formalization that is available for that study. But even this means formalization from the one point of view from which I see all such.

So I find questions asking what logic could, or should, be our primary logic bizarre. Surely choosing a logic is very little like choosing a life style; we can sit back and ponder alternative modes of comportment, but all we can ponder when it comes to logic are alternative formulations and systems, competing styles of discourse. What choice we make may matter, just as choice of a system of coordinates can matter. It can affect not only what we say, but what we see, how we see it, and what further questions we raise. But on my view, the notion of "underlying logic" serves only to cloud Professor Körner's picture. Without it his picture would seem much more promising.

Washington University

NOTE

[1] David Sanford, 'Disjunctive Predicates', *American Philosophical Quarterly* 7 (1970), 162–70.

DONALD HOCKNEY

CONCEPTUAL STRUCTURES

1. In this investigation I produce a mathematically precise theory of conceptual structures which I believe to be essential to the solution of problems arising from the phenomenon of theoretical change. Many accounts of theoretical change in the literature fail to do justice to the different kinds of theoretical change which in fact take place. This deficiency is due to a studied myopia probably having its antecedents in positivistic theories in which everything is simple. The view that theoretical change is all of one kind is an invariant through otherwise conflicting positions. It seems clear, then, that this traditional view should be rejected.

In Section I, I present a critique of Quine's notion of conceptual structures[1]. This theory is chosen for a number of reasons. It is relatively clear, important, well known (but not in the sense that it is well understood), and it admirably illustrates the importance of the notion for discussions of theoretical change. I am in sympathy with Quine's naturalism, but regard his theory of conceptual structures as an anomaly. Section II is devoted to presenting an alternative theory. In Section III I make some remarks about the nature of logic and other issues pertinent to the theory developed in Section II.

I

2. In *Two Dogmas of Empiricism*, Quine mentions such items as macroscopic physical objects, objects at the atomic level, forces, and abstract objects of various sorts as posits of a conceptual structure.[2] But he is not suggesting that rocks, electrons, nutrinos, force fields and numbers belong to conceptual structures. The ingredients intended are terms, be they singular or general, concrete or abstract, accompanied by devices for individuation and quantification.[3]

3. Since one posits an entity or is ontologically committed to it by asserting or believing a sentence, the question naturally arises as to what

sentences are included in a conceptual structure. The answer, relative to a given person, is: all the sentences believed by that person to be true.[4] Indeed, we may view the set of sentences included in a man's conceptual structure as the union of his theories, for:

> a man's theory on a given subject may be conceived, nearly enough, as the class of all those sentences, within some limited vocabulary appropriate to the desired subject matter, that he believes to be true.[5]

This notion of theory differs from many found in the literature. For example, Nagel construes theories as independent of experimental laws and thus would not treat all the sentences that a person believes to be true, relative to a given subject, as constituting his theory. Rather, he would take a sub-class of those sentences and treat those as expressing the theory. This view is based, in part, on the belief that an experimental law "retains a meaning that can be formulated independently of the theory."[6] The question of meaning invariance involved here has been discussed extensively. I will not here review arguments for or against Nagels's view. For the present, I simply wish to remark that the difference between the two views is not stipulative. It is not merely that a Nagelian theory is a subset of the set of sentences constituting a Quinean theory; the differences are reflected in doctrines of meaning, observation and theoretical change. If we speak of stipulation here, we must recognize that this is a case where stipulation is not *mere* stipulation.

4. A person's conceptual structure, then, consists of a language and the union of the theories believed by that person. Quine puts it as follows: "Language settles the sentences and what they mean; a theory adds, selectively, the assertive quality or the simulation of selected belief. A language has its grammar and semantics; a theory goes farther and asserts some of the sentences."[7] However, more remains to be said, for the relation between language and theory is crucial for a correct understanding of what Quine takes to be a conceptual structure.

5. A main feature of Quine's view on the relation of language to theory is a mutual dependence of what can be abstracted from language as the semantical component and theory itself. Quine writes:

> We learn thus to use the component words to form new sentences whose relative truth conditions are derivable. Which of these dependencies of truth value are due to meaning, or language, and which belong rather to a substantive theory that is widely shared,

is in my view a wholly unclear question. It is no mere vagueness of terminology that makes language and theory indistinguishable in this connection.[8]

Although Quine talks only of those sentences whose truth dependencies are other sentences, the view he is putting forward may be generalized. Here, it is helpful to think of a language in a straightforward way as a double, i.e., ⟨syntax, semantics⟩. The semantical element assigns an interpretation to those structures generated by the syntactic component. The generalization of Quine's account, as given above, consists in saying that it is a mistake to think of meaning as one thing and theory as another. The meaning assigned to a given sentence in a theory may be represented as a function of the semantics. However, one must recognize that the semantics itself is a function of widely shared community belief. Observation statements, so called, are not excluded. For, although the conditions under which such sentences, considered unparsed, are assigned a truth value are non-linguistic (having to do with stimulus meanings), shifts in such conditions may be due, in any case, to a shift in community wide belief. Thus sentences of any degree of observationality may be said to be theory laden in just this sense.

6. The popular opposing view which sharply contrasts semantics and theory, and thus meaning and belief, is misguided; but not simply because it depends on a notion of analyticity which Quine takes to be indefensible. Even if there should be something approximating to our intuitive demands on analyticity this would hardly settle the issue against Quine. Given that truth value dependencies may be represented as semantical functions on sentences, it by no means follows that these semantical maps are independent of shared belief. A change in belief can affect a change in meaning; and one must add that no change in meaning is a *mere* change in meaning. I have said that analyticity is not the decisive consideration here. One might be led to believe that it is because of familiar arguments for the independence of meaning relations from particular beliefs, based essentially on notions of analyticity. The writings of C. I. Lewis are representative of this type of view. Clearly, if one discredits analyticity, the grounds for the asserted independence vanish. But to stress this consideration alone is to miss the point which I regard as essential, namely that the analyticity of the sentence is not a sufficient ground for its acceptance. The doctrine of analyticity must be kept dis-

tinct from the doctrine of *a priori* knowledge. I shall return to this theme later. It is crucial for an epistemology of semantic entailment and its cousin, formal consequence.

7. To recapitulate: a person's conceptual structure, on Quine's view, consists of: a syntax, a semantics – a set of maps from sentences generated by the syntax into a specified domain, which maps depend on shared beliefs relative to some community, and the (set theoretical) union of his theories – that is, sets of sentences he believes to be true, relative to some vocabulary and some subject matter.[9]

8. I now turn to the question of individuating such conceptual structures. Here we are in for some surprises. One would naturally think that changes in reference, as a function of change in semantics, would suffice for differentiating conceptual structures. It seems an obvious view to hold given this notion of a conceptual structure. I say 'obvious' because difference in reference is just difference in posits and what is posited is crucial for Quine's notion of a conceptual structure. That matters are not so simple becomes clear once one attends to Quine's doctrines of the relativity, and the inscrutability, of reference.

9. Quine compares a language, with its predicates and auxiliary devices, to a coordinate system. In terms of this analogy, we may distinguish three theses concerning reference. The first thesis, (R_1), is as follows:

(R_1)   Questions of reference have no meaning except relative to some background language, just as questions concerning position and velocity have no sense except relative to a coordinate system.

Thus Quine says: "Relative to *it* we can and do talk meaningfully and distinctively of rabbits and parts, numbers and formulas."[10] The second thesis is that

(R_2)   Given the totality of sentences relative to a language, there is no sense in asking what the reference of any term really is, except relative to another background language, just as, given the position of a body relative to a coordinate system there is no sense in asking what its position really is, except relative to another coordinate system.

That Quine has this in mind is clear:

> It is meaningless to ask whether, in general, our terms "rabbit", "rabbit part", "number", etc., really refer respectively to rabbits, rabbits parts, numbers, etc., rather than to some ingeniously permuted denotations. It is meaningless to ask this absolutely; we can meaningfully ask it only relative to some background language.[11]

The third feature of Quine's view concerning the relativity of reference is that

(R_3) There is no privileged background language just as there is no privileged coordinate system.

Thus he says:

> When we are given position and velocity relative to a coordinate system, we can always ask in turn about the placing of origin and orientation of axes of that system of coordinates; and there is no end to the succession of further coordinate systems that could be adduced in answering the successive questions thus generated.[12]

The above theses constitute the core of Quine's doctrine of the relativity of reference. It is a semantical doctrine and has important ramifications for ontology. For example, questions as to what the objects of number theory are have no meaning except relative to an interpretation of number theory. That is, the question "What are numbers really?" makes no sense except relative to some such theory as von Neumann's set theory, Zermelo's set theory, etc. Any sentence which is true in "unreduced" number theory, e.g., '$7+2=9$', has an interpretation in either set theory. We may say that numbers are really sets, but this will only come to saying that we can interpret the objects of number theory as sets, in one set theory or another. There is no hint of ontological priority here. Remarks such as "There are really only sets" reflect an unreflective prejudice for absolute reference; and absolute reference is absolute nonsense.

10. In those cases where is no syntactic shift the story is as follows. Consider a theory T_1. The vocabulary of T_1 consists of signs such as quantifiers, truth functions, identity, singular terms and general terms. Select those sentences which are true according to the theory. Let us say that they form a set S. These truths will be truths relative to a model M_1 for T_1' (that is, T_1 minus its interpretation). Now T_1' may be given another

interpretation: some non-empty universe of objects with singular terms each assigned some member of the domain, subsets of the domain assigned to one-place predicates, and so forth. Supposing that each sentence in S comes out true under this interpretation, we have a model M_2 for T_1' resulting in a theory T_2. So far this is all orthodox. Quine's point is not simply that the model M_2 for T_1' is available, only because there is some background theory in terms of which it may be specified. It is this and more. Namely, that any question about what the objects of T_1 "really are" is relative "to some choice of a manual of translation" of T_1 into T_2. T_1 may be interpreted as T_2, but correctness of such interpretation cannot arise except relative to the selection of some such manual. This is always the case. The protest that T_2 may simply be the "containing theory" is declared empty. It is still a case of degenerate translation depending on a homophonic rule.[13]

11. I turn now to Quine's statement of his theory of conceptual difference and its grounds. He says:

We have been beaten into an outward conformity to an outward standard; and thus it is when I correlate your sentences with mine by the simple rule of phonetic correspondence, I find that the public circumstances of your affirmations and denials agree pretty well with those of my own. If I conclude that you share my sort of conceptual scheme, I am not adding a supplementary conjecture so much as spurning unfathomable distinctions; for, what further criterion of sameness of conceptual scheme can be imagined?[14]

Thus we may say: A's conceptual structure is the same as B's conceptual structure iff the public circumstances of A's affirmations and denials agree with B's. That this is a matter of degree is of no particular interest at this point. What is significant is that, on Quine's theory, this is *all* that can legitimately be meant by difference and sameness of conceptual structure. When it comes to foreigners the differences are only apparent differences. It is still fundamentally a matter of a general sentence-to-sentence correlation that results in matching up the public circumstances of the foreigner's yeas or nays with mine.[15] Further refinements are declared illegitimate on the grounds that translation is indeterminate and thus reference inscrutable. The result is that there is no difference in ontology implied.

12. So the fact of the matter is that there is no fact of the matter, aside

from stimulus meanings equated with sentences taken unparsed. This being the case, statements about relativity of reference mediated by background languages reflect no facts. More precisely, although one may say that A's theory relative to manual M commits A to such and such an ontology, this is never evidence that A believes that entities of a certain sort exist. There is no fact of the matter beyond the public circumstances of your utterances matching up with mine via a chosen manual for translation. The result is as clear as it is paradoxical. There is no information of a factual nature concerning a theory's ontology. Thus, any particular statement to the effect that the objects of a theory T are such-and-such relative to a background language, fails to convey what the objects of the theory are. Indeterminacy and inscrutability see to that. Quine's conceptual world is primitive. (Not his theory about conceptual structures.) The facts available to him are utterance strings, nods, shakes, and stimulus hits. Conceptual difference arises if you nod enough when I shake enough given the same stimulus conditions for utterances which correspond via an arbitrarily selected manual. I have no quarrel with the doctrine that reference is relative. The trouble lies with indeterminacy and the inscrutability of reference. I will argue that the only grounds for the doctrine of the inscrutability of reference are those advanced for the indeterminacy thesis and that the latter won't wash.

13. Quine has several statements of the doctrine of the indeterminacy of translation. Here I attend to his more recent statement (1970a). Translation, understood as a mapping, via a manual, by a person T, on the sentences of a given set of sentences α, the members of which are uttered by a person or persons F, into sentences forming a set $β_1$ (where $α = β_1$ may or may not hold), is indeterminate if and only if

(a) in addition to $β_1$ there are one or more distinct sets $β_2$, $β_3$, etc., compatible with the totality of F's dispositions to assent to and dissent from sentences in α, just as different physical theories may be underdetermined by all possible observations and empirically equivalent.

and

(b) There is no question as to what F "*really* believes" although T may believe some $β_i$.[16]

Inscrutability of reference is a *consequence* of indeterminacy of translation. However, Quine thinks that inscrutability of reference *stands on its own* and cites as his evidence the case of Japanese classifiers.[17] On one account, classifiers decline a neuter numeral and the declined numeral modifies an individuative term such as "ox". The other story is that the classifiers along with a mass term such as "cattle" yield "a composite individuative term", modified by the neuter numeral. Thus, on the first account the third word is a term of divided reference, whereas, in the second case, it is a mass term. His conclusion is that reference, not merely meaning, is inscrutable. "Between the two accounts of Japanese classifiers there is no question of right and wrong.... It is indeterminate in principle; there is no fact of the matter."[18] The reason why Quine thinks that inscrutability of reference stands on its own is clear. It is because both accounts of the Japanese classifier fit equally well with overall English translations of whole sentences as well as components such as "five oxen". But this is to be expected, simply because stimulus synonymy is a reliable guide for sentence pairing in cases such as this, where the sentences are high in degree of observationality. Here indeterminacy does not threaten and by the same token questions of inscrutability do not arise. However, when we turn to finer structures and query the reference of terms, it is precisely the indeterminacy of translation and nothing else that issues in the inscrutability of reference, as illustrated by the Japanese classifiers. Quine says it himself. He speaks of reference as "behaviourally inscrutable", and "indeterminate", and places the blame squarely on the choice of apparatus of individuation.[19] Any such apparatus constitutes the core of a manual of translation that goes beyond initial pairings of sentences (taken as wholes) high in degree of observationality. The grounds for the doctrine of inscrutability of reference are just those which issue in indeterminacy of translation. Quine has given no reason to believe otherwise.[20] To say that reference is inscrutable is just to say that when it comes to asking what the terms of another speaker's utterance are true of, there is "no fact of the matter". Competitive manuals of translation sort out terms in different ways. There are no grounds for preference, for saying what extensions attach to the speaker's terms.

14. I now turn to Quine's reasons for asserting indeterminacy. It is

instructive that Quine contrasts indeterminacy of translation with the underdetermined character of physical theory. He says that it is "additional." The additional element is captured in clause (b) above (§ 13). The translator might believe a given theory but there is no significance to the question of whether the foreigner really believes any particular theory. This I find astounding, given other views that Quine advances. Having dismissed propositions, Quine takes sentences to be objects of belief. Sentences belong to languages, and languages possess an apparatus for individuation. Propositions are language neutral; not sentences. If I wish to tell you what I believe about some subject matter, I will produce a string of sentences. This piece of discourse will be my theory, or partial theory. Now, the question as to what I "really believe" given my declaration, is, on Quine's view, without sense. You may rely on a homophonic rule, or permute the totality of sentences which I utter onto itself, just so long as you preserve the totality of my dispositions to verbal behaviour. There is no question of right choice of manual of translation here. Radical translation, as Quine tells us, begins at home. I may upon request "tell you again" in other or the same words. But in either case, I depend myself on a manual of translation, however degenerate. If queried further, I shall fail to say what I "really believe" simply because there is nothing in my behaviour to settle the question for my audience, and thus nothing to settle the question for myself.[21] The question as to what I really believe asked by myself or by another is a question whose "very significance" Quine must doubt, on the grounds adduced for indeterminacy of translation. Clearly, this extends to questions concerning the extension of my terms. What then of the contrast between theory and translation? There is none. On Quine's view, radical translation begins at home; and it is at home that we see the myth of the contrast.

15. Note how the above compares with the doctrine of the relativity of reference. On that doctrine, there is no unique answer to the question of what the objects of a theory really are. But this is far from claiming that amongst the non-unique answers we cannot identify an answer. One can identify the relevant coordinate system. Indeed, Quine says, "We can meaningfully ask it [what our terms refer to] only relative to some background language."[22] But what meaning is there to such a question in view of indeterminacy? The background language itself fades into

the oblivion of alternative, equally legitimate, sets of analytical hypotheses.

16. At this point it would be cheap to say that there is no indeterminacy of translation since clause (b) of the condition is violated. Indeterminacy may be viewed as extending to physical theories insofar as they are conceived as *objects of belief*. That is to say, Quine could hold that the question as to what physical theory anyone believes has no right answer – that there is no fact of the matter here, just because any attempt to answer the question is a special case of indeterminacy. That this is contrary to his stated view (as it appears in Quine, 1970a) is of little interest. What is of interest is that this contrast between physical theory and translation can not consistently be maintained, and that no conclusion should yet be drawn about the indeterminacy thesis. The course I will follow is to replace (b) with:

(b′) There is no evidence available for selecting a set of analytical hypotheses from alternative sets; but this has no parallel in physics, for here theory serves as a parameter.[23]

There is an intended ambiguity in (b′) since, in one passage, Quine contrasts the determination of a sentence as true or false within a theory with the choice of a set of analytical hypotheses,[24] and in others, the contrast is between the latter and the choice of a theory among competitive theories.[25] I will discuss these contrasts successively.

17. As evidence for the contrast (in the first sense) between physical theory and translation, Quine cites the fact that "we are always ready to wonder about the meaning of a foreigner's remark without reference to any one set of analytical hypotheses."[26] It is clear that this is regarded as support for the thesis that there are no parameters in terms of which one may select a preferred set of analytical hypotheses. But it is no evidence at all. I may wonder about the truth of a sentence uttered by a physicist in a parallel fashion. That is, I may wonder whether there is an unconceived theory in which the sentence comes out true or false. It is an idle wonder if I do nothing about it; but so is the wonder about what the foreigner means. Both wonders may or may not be cashed out in theory construction. Evidence here is not to be found in the phenomenon of idle wonder, but by an investigation of those features of physical theory and translation which are pertinent to the issue.

18. Translation begins by pairing sentences of a foreigner's language with sentences of the home language on the grounds of (approximation to) identity of stimulus meanings. The sentences treated are those high in degree of observationality and thus the pairings are subject to the usual sort of inductive uncertainty. No appeal to the apparatus of individuation which constitutes the core of a set of analytical hypotheses is required or made. Here, one has recourse to the canons of evidence within our aggregate science. Indeterminacy does not threaten such initial sentence pairings. Similarly, in physical theory, sentences high in degree of observationality do not depend for their truth on any particular theory. The conditions for assent to or dissent from such sentences are non-linguistic stimulations.[27]

19. One might object here that every sentence is "theory laden" and hence the independence from theory mentioned above is simply part of the old myth of the two-tiered structure of a scientific language. The objection is misplaced. I fully agree that observation sentences are theory laden, but assert their independence from theory in so far as verdicts to such sentences are "directly keyed to a present stimulation." This is a matter of degree, but that is of no significance for the issue at hand. It is true that what sentences turn up as observation sentences is a function of theory. Indeed, the extensions of terms which appear in observation sentences are determined, in part, by community beliefs. However, it by no means follows that verdicts on such sentences are not directly keyed to present stimulation. This is important for considerations having to do with confirmation. For confirmation of sentences high in degree of observationality need not depend on sameness of meaning, nor extensional identity (or containment). For sentences whose terms have different meanings and different extensions may not differ in verdicts under identical stimulus conditions. This is a fulcrum on which discussions of the incommensurability of theories should turn. That it has not been is due, for the most part, to an uncritical use of the notion of theory ladenness and a lack of appreciation for naturalistic epistemology. So much for the rear guard action.

20. When one ascends to sentences that are not high in degree of observationality pairings must be mediated by choice of analytical hypotheses. But here a condition of acceptability is just that the proposed set of

analytical hypotheses turn out translations which preserve the initial pairings of sentences high in degree of observationality. Here we are confronted with the empirical slack common to translation and physical theory. For, physical theory is equally underdetermined by all possible observations. Moreover, one cannot speak of the truth of such sentences independent of the theory of which they are constituents. Similarly, it makes no sense to speak of the correctness of translation for such sentences independent of the set of analytical hypotheses. Truth and correct translation are determined relative to theory and manual respectively, but meaningless otherwise.

21. The contrast between physical theory and translation enters given the underdetermined character of both by all possible observations. One must be cautious at this point, for Quine at least once draws the contrast between determining the truth of a sentence in a theory, and selecting a set of analytical hypotheses from among competitive sets.

In being able to speak of the truth of a sentence only within a more inclusive theory, one is not much hampered; for one is always working within some comfortably inclusive theory, however tentative. Truth is even overtly relative to language, in that e.g., the form of words 'Brutus killed Caesar' could, by coincidence, have unrelated uses in two languages; yet this again little hampers one's talk of truth, for one works within some language. In short, the parameters of truth stay conveniently fixed most of the time. Not so the analytical hypotheses that constitute the parameter of translation.[28]

This contrast goes nowhere. Questions about the truth of a sentence *within* a theory compare with questions of the correctness of a translation given a set of analytical hypotheses. Here there is no contrast. Theory and translation fare alike. This was the point of the last paragraph. The problem of selecting a set of analytical hypotheses compares with the problem of selecting a theory. It is here that Quine must make his case. With reference to (b'), the parameters of theory are just those canons of scientific method to be found in our "own particular aggregate science."[29]

22. Let us now suppose that we are confronted with two different sets of analytical hypotheses A_1 and A_2, both of which are empirically equivalent in the sense that they yield identical pairings of all observation sentences. Quine's point is that our aggregate science with its methodology lends no hand to the selection of a preferred manual; but that this is not so for different but empirically equivalent physical theories. This

is precisely the contrast between physical theory and translation which issues in the indeterminacy thesis and the inscrutability of reference.

23. Quine suggests that our choice between analytical hypotheses might be guided by simplicity, or that we might reject both A_1 and A_2 on the grounds that they issue in "forbiddingly circuitous and cumbersome translation rules."[30] The third possibility is that we might find that A_1 and A_2 are both "reasonably attributable." In speaking of this third possibility Quine points out that "no basis for a choice can be gained by exposing the foreigner to new physical data and noting his verbal response, since the theories... fit all possible observations equally well." There is no difference here between physical theory and translation. The three possibilities, and the comments on the third apply equally to selecting a physical theory. We are free to employ and investigate the ramifications of all utilities involved in choice of competitive theories. Manuals of translation are theories in their own right and form part of that aggregate science from which Quine draws when it comes to questions of choice between empirically equivalent, competitive physical theories. Moreover, any sceptical arguments addressing themselves to the question of utilities and truth apply equally to the problem of selecting a physical theory and a set of analytical hypotheses. Quine has offered no reason to the contrary other than the remark that one might wonder about the meaning of a foreign sentence without reference to any particular set of analytical hypotheses. But as we saw earlier, this can hardly establish the contrast between physical theory and translation that Quine draws.

24. Whatever utilities we have recourse to, there is always the possibility that they will not afford grounds for preferring one set of analytical hypotheses over another. It is with reference to this type of case that Quine says indeterminacy enters in full force.[31] But what are we to say of theories T_1 and T_2 given parallel considerations? Of what use is our aggregate science here? There is no question of right choice, for nothing in existent, accepted methodology within our aggregate ongoing science permits a choice. One may choose as he likes. The same holds true of our "rival" sets of analytical hypotheses, A_1 and A_2. Here, there is no fact of the matter. What matters however is that we rid ourselves of the myth that there is such a dichotomy be-

tween the selection of a manual of translation and a theory. The indeterminacy thesis argues for a distinction between *one* kind of scientific theory and *all* others. Quine eschews such distinctions as the analytic-synthetic, but is ready to draw a sharp distinction here. This seems odd and it is. Its oddness lies in the fact that the indeterminacy thesis is an attempt to embed the old mentalistic hypothesis that we all might have different ideas when we utter the same sentences (the ideas being the meanings) in a naturalistic theory. Quine explicity says that "sense can be made of the point be recasting it...."[32] He then goes on to speak of permuting the "sentences of a speaker's language onto itself" so as to preserve "the totality of the speaker's dispositions to verbal behavior." This old hypothesis is either understood in such a way that it is theoretically isolated and rendered untestable by definition, or in such a way that it is testable but highly improbable given present psychological theory. Thus Quine's doctrine of indeterminacy is nothing less than an abortive attempt to make an untestable or highly improbable hypothesis plausible by imbedding it in a naturalistic philosophy – a philosophy in which such hypotheses are properly shunned. The objection that Quine defends a behavioral counterpart of the old hypothesis and not the old hypothesis itself is without force on two counts. In the first place he says that *"sense can be made"* of the old hypothesis by recasting it, and secondly the behavioral counterpart is nothing but the hypothesis that translation is indeterminate. But this latter hypothesis is unsupported by arguments and, as I have urged, contrary to that hypothesis, translation, like all theories of natural science, is simply and only underdetermined by all posible evidence.

25. In view of the preceding argument the twin theories of indeterminacy of translation and inscrutability of reference are rejected; but not the theory of the relativity of reference. One can say what the objects of a theory are; but this is relative to theory; not theory neutral. Claims such as, "Numbers are really sets," are dismissed. They are as illegitimate as claims about absolute position. There are no privileged points in either space. We saw what Quine's account of conceptual difference amounted to, given the indeterminacy thesis. It needs further to be argued that the account is deficient relative to the demand that conceptual structures mirror different kinds of theoretical change. In-

determinacy intruded as a cancelation device. But dropping indeterminacy still leaves us an inadequate theory. A person's conceptual structure, at a given time, contains the union of his theories; the sets of sentences that he believes true relative to some subject matter. Thus, if I change my belief about the measurement of electrons, or my belief about the position of a coffee cup, I have in each case a change in conceptual structure. The latter should not count as a conceptual shift at all; but even if it does, there is nothing in Quine's notion of conceptual structures to mark any difference between it and the former. A change is a change, and that's the end of the story. I take it to be apparent that we require another theory of conceptual structures, and to this I now turn.

II

26. The theory to be presented finds a clue in Quine's writing on the topic of ontological reduction. I shall present the ingredients of that theory which are germane to my interests.

Consider two theories θ_1 and θ_2 whose vocabularies contain only truth functions, predicates, variables and quantifiers. Other items, such as function signs and abstraction operators are eliminable in usual ways. θ_1 is reducible to θ_2 just when: (1) there is a proxy function (not necessarily belonging to θ_1 or θ_2) whose admissible arguments are all objects of the universe of θ_1 and whose images are a subset (proper or not) of objects in the universe of θ_2; (2) Corresponding to each n-placed primitive predicate P_i in θ_1, for each n, there is an open sentence S_i of θ_2 with n free variables such that P_i of θ_1 is fulfilled by the n-tuple of *arguments* of the proxy function when and only when S_i of θ_2 is fulfilled by the n-tuple of *values* of the proxy function for those arguments.[33]

27. Proxy functions are just functions except that they may be expressed in terms of virtual classes, which amounts to talk of classes reducible to non-set theoretic items.[34] (This feature is a side benefit which, I take it, needs no further comment.) Proxy functions may or may not be one-to-one. Ontological reduction via proxy functions is structure preserving and accommodates the truths of θ_1, not merely its theorems.

28. An important feature of ontological reduction via proxy functions

is its sensitivity to the intended interpretation of the syntax of the theories considered. Essentially, we are concerned with homomorphisms and isomorphisms between such intended models. When it comes to propositional languages, ontological reduction via proxy functions does not apply. Here I consider the associated Lindenbaum algebras and homomorphisms and isomorphisms between them. Lindenbaum algebras for quantificational languages (associated with physical theories) are not considered because of insensitivity to intended interpretations. One further point needs clarification. The usual course is to construct a Lindenbaum algebra from the formal logic rather than from the language.[35] That this is but one of two available courses, may be seen as follows. Here I restrict myself to a classical propositional formal logic and a classical propositional language. The *logic* is a triple: $\text{CFL} = \langle \Gamma, T, \vdash \rangle$ where Γ is a syntax, T a set of theorems and $\vdash$ a relation on Γ defined in the usual way (the consequence relation). The *language* is a double: $\text{CPL} = \langle \Gamma, V \rangle$, where Γ is a syntax and V a set of admissible valuations on Γ specified in usual ways. A Lindenbaum algebra for CFL is constructed by defining equivalence classes of sentences belonging to Γ in CFL. First an equivalence relation '$\equiv$' is defined as: $\phi \equiv \psi$ iff $\vdash \phi \to \psi$ and $\vdash \psi \to \phi$ where '$\vdash \alpha$' is read 'α is a theorem of CFL.' The construction proceeds in familiar ways. The point is that the algebra here reflects a proof theoretic structure. This way of proceeding is to the point when one is after an algebraic completeness proof. However, it is easy to see that a Lindenbaum algebra is constructable from the *language* CPL. What follows parallels Bell and Slomson's construction of a Lindenbaum algebra on the logic. First we introduce an equivalence relation '$\equiv$' between members of the set of formulae F in Γ of CPL thus: $\phi \equiv \psi$ iff $\Vdash \phi \to \psi$ and $\Vdash \psi \to \phi$, where '$\Vdash \alpha$' is read 'α is valid in CPL.' We then proceed in the usual fashion and define equivalence classes of formulas as: $|\phi| = \{\psi \in F : \phi \equiv \psi\}$. Then we define $F/\equiv$ as: $F/\equiv \, = \{|\phi| : \phi \in F\}$. The relation $\leq$ on $F/\equiv$ is given by: $|\phi| \leq |\psi|$ iff $\Vdash \phi \to \psi$. Thus we have a Boolean algebra $\mathscr{A} = \langle F/\equiv, \leq \rangle$, i.e., a complemented distributive lattice in which $|\phi| = 1$ iff $\Vdash \phi$ and $|\phi| = 0$ iff $\Vdash \neg \phi$. This algebra is a Lindenbaum algebra. Perhaps it should be called "a Lindenbaum semantical algebra" to distinguish it from a Lindenbaum proof theoretic algebra. The Lindenbaum semantical algebra mirrors semantical properties algebraically. These are the appropriate algebras for my concerns, for it is precisely the structural features of *lan-*

guages that concern me. Hence, when I speak of Lindenbaum algebras and homomorphisms between them, I always have Lindenbaum semantical algebras in mind.

29. The method for generating conceptual structures is general. Thus it is specified for an arbitrary set of theories of natural science: $\alpha = \{\theta_1, \theta_2, \ldots \theta_n\}$ for $n \geq 2$. My primary concern is with physical theories. Methods for generating the propositional and first order languages for physical theories are available, and I do not repeat them here.[36] For my purposes, I take it that we may construct two sets in one-to-one correspondence to α. The first set, $\beta_1 = \{Q_1, Q_2, \ldots, Q_n\}$ for $n \geq 2$, consists of the quantificational language for each θ_i in α. Each Q_i has only truth fnnctions, predicates, variables and quantifiers. As before, singular terms, operations, function signs and similar devices are considered reduced to the basis. The second set, $\beta_2 = \{P_1, P_2, \ldots, P_n\}$ for $n \geq 2$, consists of the propositional language for each θ_i in α. The procedure to be described will generate families of *conceptual structures* on four levels. In general, each member of a family will group theories belonging to the same conceptual structure at that level. These families, $\mathscr{A}_1$, $\mathscr{A}_2$, $\mathscr{A}_3$, and $\mathscr{A}_4$ split naturally into two groups, the first, consisting of $\mathscr{A}_1$ and $\mathscr{A}_2$; the second of $\mathscr{A}_3$ and $\mathscr{A}_4$.

30. (a) The first level of conceptual structures is a family $\mathscr{A}_1$, the members of which consist of those theories θ_i whose corresponding Q_i's are reducible to each other *via* one-to-one proxy functions. I shall say that each member of $\mathscr{A}_1$ is a *basket*. In any given basket, each theory is structurally identical just in the sense that they are isomorphic to one another. Here ontological relativity comes into full play. To be is to be the value of a variable; but there are no preferred values.

(b) The second level of conceptual structures is a family $\mathscr{A}_2$, the members of which consist of all θ_i reduced to a given theory by proxy functions which need not be one-to-one. The reduction is mediated by the corresponding members of β_1. Thus each member of $\mathscr{A}_2$ will contain exactly one theory to which all other theories in the set are reduced. The unique member is called a *sink* and the reduced theories are called *flows*. We had no cause to speak of flows and sinks at the first level. There, isomorphism generated baskets. Here matters are different. In terms of the parallel algebraic mode we require homomorphisms not isomorphisms. Thus one set's sink may be another set's flow. Commonality of structure gives sense to these groupings, for it is just structural relations of this sort which gives sense to the notion of conceptual structure. The structural relations determine a mosaic of flows and sinks through the mechanism for imbedding. Here ontological relativity enters as well. One may prefer a sink to a flow; but as remarked above, one set's sink is another set's flow.

31. The next two families are generated from the propositional languages of the theories involved. Thus, groupings are mediated by the corresponding P_i's of β_2. In this case we have recourse to the relevant Lindenbaum algebras for each P_i. We generate a set $\beta'_2 \{\alpha_1, \alpha_2, \ldots \alpha_n\}$ for $n > 2$ in one-to-one correspondence with β_2 and thus α. We then proceed as follows.

(a) A family $\mathscr{A}_3$ is generated whose members are sets of those theories in α whose corresponding Lindenbaum algebras are isomorphic. Again we obtain baskets; but the structures are considered ontology free in the sense that the corresponding propositional languages are ontology free. I am not saying that the theories are ontology free; but rather that the groupings obtained do not reflect structures to which ontological considerations pertain. Thus, ontological relativity has no business here simply because there is nothing with which to do business.

(b) The final family $\mathscr{A}_4$ is generated in a fashion parallel to the method for obtaining $\mathscr{A}_3$, with this difference: we group on the basis of homomorphic maps into the Lindenbaum algebra corresponding to a given theory. As in the case of $\mathscr{A}_2$, we obtain sinks and flows; and again ontology plays no role.

32. In the last two paragraphs, I spoke of the members of each $\mathscr{A}_i$ containing those theories in α. This may be somewhat misleading when it comes to speaking of structural relations and ontological matters. Recall that ontological considerations are pertinent to $\mathscr{A}_1$ and $\mathscr{A}_2$, but not to $\mathscr{A}_3$ and $\mathscr{A}_4$. This is not perspicuous in our notation, but the cure is simple. As matters stand each $\mathscr{A}_i$ would have the following type of construction:

(i) $\{\{\theta_1, \theta_4, \theta_8\}, \text{(etc.)}\}$

This notation fails to display whether ontological considerations are pertinent. Notational perspicuity, relative to this demand, may be achieved by constructing $\mathscr{A}_1$ and $\mathscr{A}_2$ as consisting of sets whose members are doubles. Each double consist of the theory and its related quantificational language. Thus, considering (i) as an example of either $\mathscr{A}_1$ or $\mathscr{A}_2$ we would write:

(ii) $\{\{\langle \theta_1, Q_1 \rangle, \langle \theta_4, Q_4 \rangle, \langle \theta_8, Q_8 \rangle\}, (\text{etc.})\}$

Theories continue to be grouped as first members of such couples. The associated quantificational languages, as second members, contain the information pertinent to ontological matters, and serve as reminders that the theories, thus grouped, depend for their grouping on structural relations obtaining between the Q_i's. In the case of $\mathscr{A}_3$ and $\mathscr{A}_4$ the construction is parallel except that doubles consists of a theory and the associated propositional language. Thus, considering (i) as an example of $\mathscr{A}_3$ or $\mathscr{A}_4$ we have:

(iii) $\{\{\langle \theta_1, P_1 \rangle, \langle \theta_4, P_4 \rangle, \langle \theta_8, P_8 \rangle\} \text{ etc.}\}$

Here it is made clear that ontological considerations do not apply.

33. Conceptual change is not all of one kind and the theory here advanced reflects this view. It has been remarked that ontological considerations are relevant to conceptual differences marked by the groupings in $\mathscr{A}_1$ and $\mathscr{A}_2$ but do not pertain to $\mathscr{A}_3$ and $\mathscr{A}_4$. But more remains to be said. Theories in the same baskets in $\mathscr{A}_1$, belong to the same conceptual structure in a strong sense. Here we demand isomorphism and this reflects the intuitive idea that theories belong to the same conceptual structure just when they are mutually interpretable. Relative to this demand we think of a conceptual shift as a move from a theory θ_1 to a theory θ_2 where either of the two theories exhibit differences due to difference in universe or the unavailability of a proxy function. While $\mathscr{A}_3$ contains baskets, any given basket in $\mathscr{A}_3$, in general, will contain all the theories in some basket in $\mathscr{A}_1$ *and more*. For, when it comes to these groupings the demand for isomorphism is not sensitive to matters of ontology reflected in the intended models associated with each quantificational language of a theory. The different groupings mediated by propositional languages reflect deeper conceptual shifts re-

lative to the demand for isomorphism. Thus, two theories may belong to different conceptual structures in $\mathscr{A}_1$, but not in $\mathscr{A}_3$. In $\mathscr{A}_3$ the structures are gross, while in $\mathscr{A}_1$ finer difference are marked. Analogous comments apply to $\mathscr{A}_2$ and $\mathscr{A}_4$, but there are further differences, since we demand only the relevant homomorphisms as a principle for grouping. Here the intuitive idea is that theories imbeddable in a yet richer theory form a conceptual structure, and that these imbeddings form a mosaic of flows and sinks. Theories which stand out because of lack of "connection" with other theories will turn up as singletons in $\mathscr{A}_2$, while those which have been extended will be grouped together. The theories grouped in different sets in $\mathscr{A}_4$ reflect the deepest conceptual differences while those grouped together reflect the loosest conceptual similarities. Conceptual difference at this level marks a theoretical change of the most drastic sort. On the view advanced here there are different ways of marking conceptual shifts. No one family of the four generated is a preferred way of grouping theories. The sense of conceptual change and similarity differs from family to family, from the finest groupings in $\mathscr{A}_1$ to the most liberal in $\mathscr{A}_4$. Each way of grouping theories answers to a well defined interest and together they provide us with ways of talking about theoretical change in a precise fashion.

34. An example is called for. The comparison between classical and quantum mechanics will serve. Here I draw on the work of Kochen and Specker. Briefly, they show the following. The algebraic structure of idempotent magnitudes represented by projection operators is the relevant algebraic space for quantum mechanics. This forms a partial Boolean algebra, that is, a partially ordered set with a (compatibility) relation which is both reflexive and symmetric. There is a one-to-one correspondence between the idempotent magnitudes, subspaces in Hilbert space, and the set of equivalence classes of theoretical sentences to the effect that the value of a magnitude belongs to a borel set of reals. A theoretical sentence of the propositional language for quantum mechanics is true iff the value of the magnitude is 1 and false iff the value is 0. The Lindenbaum algebra for the language of quantum mechanics is not imbeddable in a Boolean algebra. Since a Boolean algebra is the Lindenbaum algebra for a classical propositional language and characterizes the phase space of classical mechanics, it follows that quantum mechanics is not a flow

into the sink of classical mechanics. This is equivalent to the Kochen and Specker result. There is no set in $\mathscr{A}_3$ nor in $\mathscr{A}_4$ which contains both classical and quantum mechanics.[38]

35. Although it is not my intention to address myself to languages independent of physical theories, the method for generating conceptual structures may be applied here. I mention one case. The controversy between advocates of intuitionistic logic and advocates of classical logic is often said to involve deep conceptual differences. There are differences, but not all should be characterized as deep conceptual differences in terms of the contrast between intuitionistic logic and classical logic. For example, Brouwer's thesis that foundational studies in mathematics are intimately connected with a peculiar activity of the mind by which there is a clear apprehension of what it has constructed stands in opposition to Frege's claim that such mentalistic concerns should play no part in foundational studies. This might be a deep conceptual difference and it may be mirrored in opposing psychological theories concerning mathematical thinking. However, there is no reason to suppose that the structural relations between these psychological theories parallel the relations between an intuitionistic propositional language and a classical propositional language. With reference to the latter, the intuitionistic propositional language is a flow into the sink of the classical propositional language. This is so because there is a homomorphic map from the pseudo Boolean algebra of the intuitionistic language into the algebra for the classical language.[39] However, because of lack of isomorphism they are in different baskets in the family $\mathscr{A}_3$.

III

36. The remainder of this paper is concerned with a few problems on which the notion of conceptual structures may shed some light. Philosophers, for the most part, have regarded logic as *a priori*. The world of events in space-time is one thing and logic another. Our beliefs about the working of nature might shift, but our logic remains eternal. The doctrine takes many forms familiar in the history of philosophical thought. Quine shattered our confidence in this in his classic paper *Two Dogmas of Empiricism*. We are now in a position to see things even more clearly. Quine was of the opinion that there **is** no sense to the notion of analyticity. He

was anxious to rid us of its charm because, I conjecture, it seemed to go hand in hand with the doctrine of *a priori* knowledge. Discredit analyticity, and *a priori* knowledge has no ground. Some philosophers slipped back to the synthetic *a priori*, but most remained unconvinced of that Kantian caper. On my view, it makes good sense to retain a doctrine of analyticity relative to a language, in terms of admissible valuations defined on a syntax. I can make no global sense of the notion of analyticity. It has no place in natural languages, for the reason that there is no mathematically precise natural language, or rather formulation of one. Having said this, I call your attention to two points; and the example of quantum mechanics should make these clear. First, the languages for each theory θ_i are generated from the event space of the theory. The admissible valuations are grounded in physical theory. Thus, that a sentence is analytic for the language of quantum mechanics records the fact that the sentence in question is invariant in the theory through all transformations. It is the invariants of a theory that display its conceptual guts. Seen in this guise, analytic sentences are hardly trivial. The triviality imputed to analytic statements results from a trivial view of logic. Secondly, that a sentence S_1 is analytic in a language L_1 of a theory θ_1, is not sufficient grounds for its acceptance. There may be grounds for preferring another theory θ_2, and the language L_2 of that theory may be such that, although S_1 belongs to the sytax of L_2, it is not true in L_2. Hence, my reasons for rejecting or accepting analytic sentences, so conceived, are just my reasons for accepting or rejecting theories. On the other hand, given the amount of semantic slack in our theories, it seems silly to say that logic is empirical. Logic is no more or less empirical than physical theory itself.

37. It is a myth of long standing that there is one proper logic. Quine himself has been attracted to it. No logic is privileged. The fact that I employ a Boolean logic in arguments contained in this paper lends no comfort to those who would seek its primacy. I have no reason to think that it is not a logic appropriate for discussing these issues. However, it is not an appropriate logic for reasoning about micro events. Logical pluralism is as viable as conceptual difference. I repeat: There is no one preferred logic. There are only logics appropriate to the theories within the mosaic of theories that constitute our on-going aggregate science.

Here one must be cautious. I am not advocating conventionalism but *logical pluralism*. Conventionalism is *a priorism* gone mod. What is important on my account is the structure of the *language* of the event space of a theory. A formal logic which is argument and statement sound for that language is an appropriate formal logic for the theory. As I see it, the acceptability of a formal logical system is under the control of the *language* generated from the event space of the physical theory in question.

38. Logic, as I conceive it, is more fundamental than geometry. This is reflected by the fact that the families of conceptual structures sort out into two groups. $\mathscr{A}_1$ and $\mathscr{A}_2$ are relevant to ontological questions; $\mathscr{A}_3$ and $\mathscr{A}_4$ are not. Thus it is significant that space-time predicates such as 'acceleration' gain ontological status within $\mathscr{A}_1$ and $\mathscr{A}_2$ but that this drops out at the propositional level captured by $\mathscr{A}_3$ and $\mathscr{A}_4$. Here the sorting is less fine, relying as it does on the structure of propositional languages. We end up with Boolean and non-Boolean structures in different conceptual sets, free of consideration of extensions of terms. Here, the most fundamental structures are revealed and they are gross logical structures. Geometrical considerations depend on finer structures revealed in $\mathscr{A}_1$ and $\mathscr{A}_2$.

39. The vast and rapidly growing literature on quantum mechanics is particularly fascinating when viewed as a debate about the nature of logic. It seems clear that some researchers have an *a priori* preference for a "Boolean world." Thus we have the ignorance interpretation of quantum mechanics, the search for workable hidden variable theories and the denial of any possible consistent realistic interpretation of quantum mechanics. I am not suggesting that research programmes dedicated to the construction of a new quantum theory which is Boolean are misguided. What I am calling attention to is a belief that only such a construction can be "the truth." This attitude towards nature, and our understanding of it, would appear to be an intrusion of *a priorism* into an activity whose history is the history of purging *a priori* considerations from its endeavour.

40. I conclude with some remarks about investigations of "the ordinary man's conceptual scheme." It is far from clear just what "conceptual analysis", so typical of Anglo-Saxon philosophers who more or less

identify themselves with Wittgenstein and/or Austin, comes to. However, this much does seem clear. Often an attempt is made to describe "our conceptual scheme" in terms of the English language. But when one reflects on natural languages it becomes apparent that we are confronted with enormous difficulties. For natural languages – especially those with such diverse historical antecedents such as English – hardly can be said to embody a particular theory. We have instead a mosaic of quasi theories. Thus, there is little sense to the notion of "our conceptual scheme." It was the genius of Wittgenstein to see this point. He was convinced that his followers would misunderstand him and proved himself prophet. "Many language games are played." One might at best try to describe some of them. They are far from neat. Such descriptions yield some illumination about how some people think some of the time – perhaps.

41. I have said nothing about a host of problems relevant to the notion of a conceptual structure advanced here. Indeed, I have painted with a large brush on a small canvas. I can only hope that the brush is not too large and the canvas too small.[40]

The University of Western Ontario

NOTES

[1] Most authors, Quine included, use the terminology 'conceptual schemes'. I prefer 'conceptual structures' to stress that conceptual similarities and differences depend on structural considerations.

[2] Quine (1963c), p. 44–45.
[3] Quine (1963a), pp. 77–79.
[4] Quine (1960), pp. 4, 24.
[5] Davidson and Hintikka (1969), p. 309.
[6] Nagel (1961), p. 86.
[7] Davidson and Hintikka (1969), p. 309.
[8] Davidson and Hintikka (1969), p. 310.
[9] On my view, ingredients of the syntax itself are under theoretical constraint. This is evident in neologism and the introduction of new sentence forms.
[10] Quine (1969a), p. 48.
[11] Quine (1969a), p. 48.
[12] Quine (1969a), p. 49.
[13] Cf. Quine (1969a), pp. 53–55.
[14] Quine (1969b), p. 5.
[15] The differentiating property consists of stimulus meanings associated with sentences taken as wholes. For Quine's definition of 'stimulus meaning' see Quine (1960), pp. 32ff.

[16] See Quine (1970a), pp. 179–81.
[17] See Quine (1969a), pp. 35–37.
[18] Quine (1969a), pp. 37–38.
[19] Quine (1969a), p. 35.
[20] In Quine (1969a), pp. 41–43, it is argued that inscrutability "persists in a subtle form even if we accept identity and the rest of the apparatus of individuation as fixed and settled; even indeed, if we forsake radical translation and think only of English." But forsaking radical translation is accepting the apparatus of individuation. Moreover, the argument given supports the relativity of reference, not the inscrutability of terms.
[21] For Quine, there is no private language. To postulate one would go counter to the thesis that there is nothing in meaning that is not in behavior. Cf. Quine (1969a), pp. 27–29.
[22] Quine (1969a), p. 48.
[23] See Quine (1960), pp. 24–25; pp. 75–76; Davidson and Hintikka (1969), p. 303; Quine (1970b), p. 10ff.
[24] See Quine (1960), pp. 75–76.
[25] See Davidson and Hintikka (1969), p. 303 and Quine (1960), pp. 24–25.
[26] Quine (1960), p. 76.
[27] See Quine (1960), p. 44.
[28] Quine (1960), pp. 75–76.
[29] Quine (1960, p. 24. Cf. Davidson and Hintikka (1969), p. 303 and Quine (1970a), p. 180.
[30] Quine (1970a), p. 180.
[31] Quine (1970a), pp. 180–81. Here I am ignoring the fact that Quine states the indeterminacy thesis in terms of what the foreigner really believes. This form of the doctrine has been treated above.
[32] Quine (1960), p. 27. Commenting on Stenius, Quine says this passage "tells my purpose". Davidson and Hintikka (1969), p. 299.
[33] See Quine (1966), p. 305. The above is a paraphrase of Quine's statement. It suffers from lack of mention of the case where objects may be dropped from the universe of a theory without falsifying truths, but not in virtue of a proxy function. (See Quine (1969a), p. 68.) Throughout, I assume that all such 'reductions' are carried out on the condition that the class of discarded objects is specifiable in the background language.
[34] See Quine (1963b), §2.
[35] See Bell and Slomson (1969), p. 41ff.
[36] The reader should consult the following: Bub (1973); Finkelstein (1969); Kochen and Specker (1967); Putnam (1969). A brief example of constructing a language from a theory is provided in § 34 of this essay.
[37] The exposition given here is woefully incomplete. The papers cited in note 36 should be consulted for a detailed account.
[38] Kochen and Specker (1967) prove that for a partial Boolean Algebra H:

"1. H is imeddable into a Boolean Algebra if and only if, for every classical tautology of the form $\phi \equiv \psi$, $\phi = \psi$ is valid in H.
2. H is weakly imbeddable into a Boolean Algebra if and only if every classical tautology ϕ is valid in H.
3. H may be mapped homomorphically into a Boolean Algebra if and only if every classical tautology ϕ is not refutable in H." (p. 84)

(A weak imbedding is "a homomorphism which is an imbedding on Boolean subalgebras of H".) From the theorem that the algebra of quantum mechanical observables on Hilbert space cannot be imbedded in a Boolean algebra and (1)–(3) above it follows that there is a classical tautology which is false under an admissable substitution of quantum mechanical propositions for its variables. Some authors see this failure as the distinction between quantum and classical mechanics. (e.g. Putnam, 1969.) This is misleading because it does not reveal the structural differences that distinguish the two theories. The fact that there are classical tautologies which are quantum mechanically invalid is a consequence of the non-existence of a homomorphism between the algebra of quantum events and a Boolean algebra.

[39] See Raisowa and Sikorski (1968), p. 395ff.

[40] It is obvious that I owe much to W. V. Quine. I enthusiastically share his naturalism, however much I disagree in detail. I wish to express thanks to my colleague Jeffrey Bub with whom I have had many hours of fruitful conversation concerning the ideas developed in this paper. John Nicholas, Glenn Pearce, Patrick Maynard and William Demopolous have read this essay and offered a number of helpful suggestions.

BIBLIOGRAPHY

Bell, J. L. and Slomson, A. B., 1969, *Models and Ultraproducts*, Amsterdam.

Bub, Jeffrey, 1973, 'On the Completeness of Quantum Mechanics', in *Contemporary Research in the Foundations of Quantum Theory* (ed. by C. Hooker), D. Reidel, Dordrecht, p. 1.

Davidson, Donald and Hintikka, Jaakko (eds.), 1969, *Words and Objections Essays: on the Work of W. V. Quine*, D. Reidel, Dordrecht.

Finkelstein, David, 1969, 'Matter, Space and Logic' in *Boston Studies in the Philosophy of Science*, Vol. V (ed. by R. Cohen and M. Wartofsky), D. Reidel, Dordrecht, pp. 199–215.

Kochen, S. and Specker, E. P., 1967, 'The Problem of Hidden Variables in Quantum Mechanics', *Journal of Mathematics and Mechanics* **17**, 59–87.

Nagel, E., 1961, *The Structure of Science*, New York.

Putnam, H., 1969, 'Is Logic Empirical?' in *Boston Studies in the Philosophy of Science*, Vol. V (ed. by R. Cohen and M. Wartofsky), D. Reidel, Dordrecht, pp. 216–41.

Quine, W. V., 1963a, 'Identity, Ostension and Hypostasis', in *From a Logical Point of View*, (ed. by W. V. Quine), 2nd ed., New York pp. 65–79.

Quine, W. V., 1970a, 'On the Reasons for Indeterminacy of Translation', *The Journal of Philosophy* **67**, 179–81.

Quine, W. V., 1966, 'Ontological Reduction and the World of Numbers', in *The Ways of Paradox and Other Essays* (ed. W. V. Quine), New York, pp. 199–207.

Quine, W. V., 1969a, 'Ontological Relativity', in *Ontological Relativity and Other Essays* (ed. by W. V. Quine), New York, pp. 26–68.

Quine, W. V., 1970b, 'Philosophical Progress in Language Theory', in *Language, Belief and Metaphysics* (ed. by H. Keifer and M. Munitz), Albany, pp. 3–18.

Quine, W. V., 1963b, *Set Theory and Its Logic*, Cambridge, Mass.

Quine, W. V., 1969b, 'Speaking of Objects', in *Ontological Relativity and Other Essays* (ed. by W. V. Quine), New York, pp. 1–25.

Quine, W. V., 1963c, 'Two Dogmas of Empiricism', in *From a Logical Point of View* (ed. by W. V. Quine), 2nd ed., New York, pp. 20–46.

Quine, W. V., 1960, *Word and Object*, Cambridge, Mass.

Raisowa, H. and Sikorski, R., 1968, *The Mathematics of Metamathematics*, 2nd ed., Warsaw.

NORETTA KOERTGE

THEORY CHANGE IN SCIENCE*

> ... the crucial point that should never be forgotten in the history of ideas...: one may have been influenced profoundly by others and yet be strikingly original and even revolutionary. What makes the study of history fascinating is among other things, the perception of discontinuity in the context of continuity. The historically ignorant believe in absolute novelty; those with a smattering of history are apt to believe in no novelty at all: they are blinded by the discovery of similarities. Beyond that, however, lies the discovery of small, but sometimes crucial, differences.
>
> KAUFMANN, *Faith of a Heretic*

Nowadays theories about science change even more rapidly than science itself. Not too long ago nearly everyone would have agreed that the primary method of criticizing scientific theories was falsification; (the disagreements were over what positive things could be said about theories which survived testing). Then Kuhn (1962) discovered normal science and Agassi (1966) argued that Boyle's Rule was dogmatic (why should one always keep the experimental result and drop the theory?) and so it was suggested that the method of science was the removal of inconsistencies, preferably not by *ad hoc* stratagems. However Feyerabend (1962) soon showed that the demand for consistency was much too stringent since most progressive new scientific theories were inconsistent with the best theories and observational evidence available at the time. Lakatos (1970) then proposed a sort of Pollyanna theory of science – scientific research programs should be evaluated by counting only their successes and ignoring their failures. But Feyerabend (1970) argued that even this standard was too restrictive and claimed that the methods of science were counter-induction, proliferation, and "anything goes."

Faced with the choice between Lakatos' instrumentalistic opportunism and Feyerabend's post-rationalistic anarchism, anyone who wants to make a further proposal concerning the methodology of science has no alternative but to become a conservative! And indeed I will claim that the aim of science is true, coherent, explanatory theories, that science progresses by removing inconsistencies, and that certain patterns of criticism which look remarkably like good old-fashioned Popperian falsifications play an important role in the development of science.

This paper attempts to provide both an idealized description of the way science has in fact developed and a rational analysis of some of the typical methodological decisions which scientists make. I also try to offer a satisfactory account of the kinds of historical phenomena to which Feyerabend, Lakatos, and Kuhn have drawn our attention.

My first problem is suggested by the quote from Walter Kaufmann (1963). Can one give an account of "discontinuity in the context of continuity" which will be an accurate description of the relationship between successive scientific theories and yet be non-trivial? I suggest that Post's General Correspondence Principle provides a good approach to this problem, but criticize some details of his formulation. I next propose a theory concerning the role of what I call "preferred statements" in scientific theories. This theory is intended to give an account of both the discontinuities in scientific development and the long periods of theoretical conservativism. However, the proposed description of the development of science raises serious questions about the rationality of science: (i) Can good arguments be given for declaring a particular set of statements to be preferred ones? (In short, is "normal science" ever rational?) (ii) Can good reasons be given for revising one's set of preferred statements? (Is the process of theory change a rational one?)

My answer to both questions is a qualified "yes". I then describe methods for choosing, criticizing and changing preferred statements and claim that these decisions are rational ones.

The Whewellian approach of using the history of science to investigate the methods and logic of science has many advantages. Many of the most interesting questions concerning good scientific practice only emerge when one moves away from white swans, balls in urns, grue emeralds, and non-black non-ravens and begins to look at the sophisticated problems and comprehensive, highly-developed theories which scientists actually investigate.

However, there are also dangers in using an historical approach. One may forget that the practice of scientists is at least as fallible as their theories and attempt to make one's theory about science so weak as to cover everything. Also the very complexity of the problems-theories-evidence situation in any historical episode may tend to obscure the simple epistemological and methodological principles which provide an explanation of the rationality of science. Finally, anyone who attempts to provide a comprehensive descriptive theory of scientific growth is always likely to be misunderstood as suggesting that the development of science obeys inexorable historical laws.[1] Of course I do not believe that it is inevitable either that we succeed in improving our current theories or that we continue to do science. Institutionalized science, even when it becomes part of the establishment, is a fragile enterprise, one which is strongly influenced not only by human decisions and values, but also by our theories about it. One only hopes that the scientific tradition is strong enough to survive both the attempts of philosophers to understand it and those of government to exploit it.

I. THE PROBLEM OF CHARACTERIZING THE RELATIONSHIP BETWEEN SUCCESSIVE SCIENTIFIC THEORIES

Feyerabend began his famous paper, 'Explanation, Reduction and Empiricism' (1962), with a criticism of what he called *the deducibility requirement*, viz. the requirement that every new theory have as a deductive consequence the less general theory which preceded it. For some time many philosophers and scientists had not believed that there was usually a *strict* deducibility relationship between successive scientific theories. Duhem and Popper had stressed the fact that Newton's theory was strictly inconsistent with Galileo's and Kepler's laws; physicists were very aware of what they called *the correspondence relationship* between various modern theories and the classical theories which were now seen to be only "approximately true", and that only in certain limiting cases.[2]

However, Feyerabend argued that these approximate numerical similarities between theories were of little interest because they failed to reflect the radical conceptual shifts which accompanied any revolutionary changes within the basic theoretical framework of a science. Rather than enter into the debate over the contrast between Newtonian and Einsteinian conceptions of mass, I would like to introduce a new example which is intended to illustrate one of Feyerabend's points, namely that there are sometimes radical changes in ontology as science progresses. However, we shall see that even in this case, which involves a dramatic change in ontology, there are, nevertheless, striking resemblances between the rejected theory and the new one which replaces it. This example is also chosen to show that the phenomenon of correspondence is not limited to recent theories in physics. Later I shall argue that it is a quite general aspect of the growth of science.

A. *An Historical Example of Theory Change*

Let us analyze the relationship between the force of the vacuum theory which was first introduced by Galileo and the atmospheric pressure theory of Pascal which replaced it.[3]

According to Galileo's theory, any bit of void space exerts a contracting force on the surrounding matter. This force is a constant, it is independent of the volume, surface area or shape of the vacuum, and it tends to pull things towards the center of the evacuated space. From this theory Galileo and later workers, such as Torricelli and the early Pascal, were able not only to explain the already observed behavior of siphons and lift pumps, but also to predict the behavior of a barometer before one was actually constructed. The theory could also explain why the maximum height of liquid lifted by a siphon, barometer, or lift pump was independent of the diameter of the tubing used in its construction. And once the value of the force of the vacuum was measured by using a liquid of known density, such as water, one could predict the height to which a liquid of a different density, such as mercury, could be lifted in such devices.

I will assume a familiarity with Pascal's atmospheric pressure theory which replaced the force of the vacuum theory and now turn to the question of the relationship between the two theories. In this example there would seem to be no possibility of deriving even an approximate version of *all* of Galileo's theory from Pascal's. Some of the existential claims of the force of the vacuum theory are false. Pascal recognized that there are no forces of the sort required acting within a vacuum.

But might not one argue that the expression 'force of the vacuum' should be interpreted by means of an operational definition or some similar device? Following this approach one would claim that the "empirical cash value" of the phrase 'force of the vacuum', i.e. the sets of pointer readings and experimental apparatus which elicited its use, overlapped with the "cash value" of a phrase in the new theory, 'atmospheric pressure', so that in many situations they could have been used interchangeably. One could also point out the formal similarities in the deductive explanations given within the force of the vacuum theory and the atmospheric pressure theory.

I agree that one can set up partial correspondences between the ways in which these terms in the two theories were used and that there is an overlap between the sets of gross objects and situations which elicited the terms. However, this should not mislead us into thinking that the terms 'force of the vacuum' and 'atmospheric pressure' were intended to describe the same thing. Even without debating the merits of the theory of meaning implicit in the above suggestion, we might all agree that one of the most important properties of an entity is its location (if it has one) and that if the denotata of two terms being used in paradigmatic situations are not in the same place, then the terms have different senses. If one were asked to indicate the location of the force of the vacuum and where it was acting, one's finger would not point to the same spot as if one were asked where the force of the atmospheric pressure was acting. After the fact, one might claim that the exact location of the force of the vacuum was scientifically irrelevant. But without the benefit of the new theory one would have no idea as to what the correct positivistic interpretation of the term should be.

I conclude that Galileo's theory was quite mistaken in some of its claims. Nevertheless, we also note that there are many statements derivable from it which are also derivable in the new theory. These shared statements include low level generalizations about the behavior of siphons of a particular construction. There are also statements which are seen to be approximately true within a certain domain which is specified by the new theory. For example, the height to which a lift pump can raise water *is* fairly constant if the pump is near sea level. Also derivable are higher level generalizations about how the behavior of siphons, barometers, or lift pumps is affected by the density of the liquid to be transported. If we move to an even higher level of generality we see that the old theory had succeeded in correctly positing connections between phenomena which had previously been considered to be quite unrelated. For example, the old false theory correctly asserts that the force holding up the barometer column is the same as the force acting in a siphon or lift pump.

All in all it is impressive that the truth content, or the "approximate truth content", as Fine (1971) would describe it, of a theory which was based on non-existent entities should have been so extensive! And this is not an isolated example of serendipity – one can give similar analyses of the phlogiston theory of combustion, calcination, reduction and nitration, the caloric theory of heat, and Spemann's organizer theory.

How shall we describe the striking resemblances between the explanatory structures of successive scientific theories? Can one make any generalizations about which parts of the old theory will be incorporated into the new?

B. *The General Correspondence Principle*

Many scientists and philosophers have held views similar to the one I shall now describe, but Post (1971) has given the most complete description of this relation between scientific theories and made the strongest claims about its universality. The position which he calls the General Correspondence Principle (GCP) is based on a historical generalization of roughly the following form:[4]

If S and L are well-confirmed theories in the history of science and L is a successor of S, then an approximate or qualified version of some subset of the statements of S (call the subset S^*) is derivable in L.

There are many ways in which one might wish to make this formulation more precise, e.g., what do we mean by saying L is a "successor" of S? Post's discussion clarifies many of these issues. But here I want to deal only with these important problems: How is S^* related to S? How is S^* related to L?

1. *The Problem of Demarcating S^**. The above generalization about the development of science would be very interesting if one could give a method of determining which statements in S belong to the class S^* which is independent of the content of L. Can one specify beforehand (i.e., without looking at L) which subset of the statements derivable in the old theory will be retained (in some cor- rected or qualified form) by the new theory? Or can one at least order statements in a theory along a most-likely-to-survive scale such that one predicts that S_{k+1} will be retained only if S_k, S_{k-1}, etc. are?

There are various criteria which seem to be plausible ones to use in demarcating S^* from the rest of S. However, I shall argue that none of them completely succeeds in handling our historical example although taken together they provide a rough guide as to which parts of the old theory we should expect to be retained by the successor theory.

(a) *Ordering by means of the observational-theoretical character of the concepts in the statements.* One suggestion relies on there being a clear distinction between observational and theoretical terms. Thus Duhem (1914) claimed that what he called "explanatory theories", i.e., theories which try to describe the world lying behind appearances, were bound to lead one into error and that only "representative theories", as he called them, could be expected to provide "natural classifications". Neither Duhem not the positivists ever produced a satisfactory absolute demarcation between theories and laws or between theoretical terms and observational terms. However, we might hope to find a useful relative distinction between the terms used within a given system in a given historical context. The claim would then be that all well-tested statements in the theory which are couched in terms having a certain minimum degree of observational character should be expected to be retained in any new theory.

However it can be plausibly argued that this attempt to order the statements of the old theory fails in our historical example. At the time, the statement that there was a force of the vacuum acting at x_1, y_1, z_1, t_1 would have been regarded as a relatively observational one. It was thought that the force of the vacuum could be directly experienced – who has not tried to inflate bellows with the intake valve closed, or felt his flesh pulled by a cupping glass, or played with a perfume bottle by sucking the air out of it and fastening it on his tongue? The predicate 'is a barometer' would have been considered a less directly observational one. To decide whether a given bit of apparatus was a barometer, one would have had to check whether the tubing was airtight or whether a transparent glass plate was holding up the mercury. In addition, although everybody at the time believed that nature abhorred a vacuum in some sense and so were ready to accept that there were forces directed towards places where matter was rarified, not everone was ready to admit that a macroscopic vacuum could be actualized. To the extent that calling something a barometer implied that it contained a vacuum, the term was laden with a highly speculative theory.

Although the criterion is not a very clear one and hence there can be disagreements concerning its application, I would tend to conclude that 'force of the vacuum' is more observational than 'barometer'. But none of the statements retained in Pascal's theory contain the expression 'force of the vacuum' while some do use the term 'barometer'. So the proposed criterion fails to order the statements such that we can predict which ones are most likely to be preserved within the new theory.

(b) *Ordering by means of the degree of universality of the concepts in the statements.* A second ordering is suggested by the quotation in the note below in which Popper speaks of degrees of universality.[5] Also in our description of the force of the vacuum theory we spoke of "low level" generalizations about the behavior of all waters barometers and "higher level" ones about the behavior of barometers filled with any liquid of known specific gravity.

However, if we wish to produce an order according to expected survival, this criterion is unsatisfactory. Often statements of a relatively high degree of universality are retained (e.g., "All columns of liquid standing in tubes are acted on by forces equal in magnitude and opposite in direction to the weight of the column of liquid") while those of a lower level are not (e.g., "All columns of liquid standing in barometers are acted on by the attractive force of the vacuum").

One might object to the above example by arguing that the more universal statement about balanced forces belonged to an auxiliary theory, viz. statics, and that it should not be considered to be part of the S theory in question. However, there seem to be many cases of theory change in which many of the most universal laws, such as conservation laws, are preserved.

(c) *Ordering by degree of confirmation.* The most promising criterion for selecting out the statements in S which are most likely to be retained in the new theory is suggested by Post (1971, p. 228), viz. the degree of confirmation of the statements.[6] Once again the criterion cannot easily be made precise and we must rely on informal judgments about relative degrees of confirmation or corroboration. Also there are serious problems in assessing the degree of confirmation of separate statements as opposed to theories. Nevertheless, this suggestion is attractive because it would seem correctly to predict that certain very high level principles such as conservation laws are likely to be retained. Also, as Post points out, sometimes one can even use confirmation results to make finer analyses of the probable fate of an S statement in the L theory. If a law has been extensively tested in a certain domain, but not in others, then we will not be surprised if the law is qualified within the new theory so as to apply only to the domain where it was originally successful.

However, in our example the confirmation criterion fails to give a correct ordering of the statements in the force of the vacuum theory. Consider the statement: "If the only unbalanced force acting on a column of liquid is the force of the vacuum, the liquid will rise until its own weight is in equilibrium with the force of the vacuum." This statement is not retained in the new theory, but surely in the heyday of the old theory it would have been given quite a high degree of corroboration. It gave a quantitatively correct prediction of a qualitatively new phenomenon (the barometer). It also correctly predicted the relative heights of water and mercury barometers. One might try to argue that this statement is not as highly corroborated as a corresponding one which does not mention the force of the vacuum. But to propose this content-decreasing manoeuvre is to cut the prediction off from the theory which made it plausible in the first place.[7]

(d) *An approximate ordering.* Given the crudeness of the proposed criteria, it is perhaps not surprising that we have not arrived at a foolproof method of selecting out the parts of the old theory which are most likely to be retained. However, I think there are deeper, epistemological reasons for this failure. Often one can give a stronger argument for the claim that a certain *group* of our everyday observation reports are veridical than one can for the claim that a *particular* observation experience is not an illusion. Likewise, one can have reasons to claim that a particular theory has a good deal of truth content without being able to make a very reliable estimate of exactly which statements are true.

I conclude that by using a criterion such as degree of corroboration one can make a rough relative estimate of the future fate of various statements in a theory. Certain parts of the theory may obviously be speculative and untested. Other parts may have survived extensive and severe testing. For example, although there were repeated attempts to correlate intense color with phlogiston content (many combustible organic materials are also highly colored), even the early phlogistonists, such as Stahl, were very aware that this was a weak aspect of the theory. Kunckel investigated a whole series of color changes involving various complexes of copper and argued that phlogiston (or "principle of sulfur", as he calls it) was not involved. Here is an example where the degree of confirmation criterion would have predicted correctly that the phlogiston theory of color would not be retained in an improved theory.[8]

Post has also stressed that our predictions about which parts of the S theory will be retained are strongly guided by coherence considerations. As we have seen in our examples, whole chunks of what might be called the explanatory structure of S are carried over into the new theory. Classification systems are retained (perhaps with minor modifications); few cases of constant conjunction are disrupted. Phenomena which were previously given a unified explanation are still seen as being similar even though the details of the explanatory law may be changed.

2. *The relationship of S^* to L.* We have already seen that in the general case S^* itself is not derivable in L because the new theory corrects or qualifies even the best corroborated claims of the old theory. Let us call the qualified version of S^* which is so derivable S^*_Q. We would not expect to be able to predict the qualification needed before the advent of the new theory. Nevertheless, there are some interesting general claims which can be made about Q, the qualification introduced by L which restricts the domain to which the S-theory statement truly applies and which specifies the degree of approximation to which the original statement is correct.[9] First (and this is perhaps surprising), these Q conditions are not *ad hoc* in any sense. For example, they never consist of just a list of exceptions to the old law. Quite the contrary – they are linguistically simple and limit the domain of application of the old theory's statements by using categories which are easily and naturally expressible within the language of the new theory. Also the specification of the degree of approximation to which the old theory is correct can be calculated using the new theory – one does not rely on a mere compilation of actual measurements.

Whenever a "natural" Q condition such as I have described above exists, and when many of the well-corroborated generalizations of S are in S^*, we shall say that L and S stand in a correspondence relation or that L corresponds to S^* in cases satisfying Q. The General Correspondence Principle is the claim that nearly all pairs of successive theories in the history of science stand in a correspondence relation and that in those cases where there is no correspondence relation to begin with, the new theory will be developed in such a way that it comes more nearly into correspondence with the old.

I will briefly discuss one historical example of this process of "completing the correspondence" between two theories. After the discovery of isotopes, it was necessary to revise Mendeleev's Periodic Law which claimed that the list of elements ordered according to their atomic weights exhibits the periodic properties associated with the Periodic Table. According to the new theory, it was the order by atomic *number* which was significant. However, at the time there was no very satisfying answer as to why the orders were so nearly the same. (In a list of approximately 100 elements there are only three reversed pairs.) For example, there are about 50 pairs of adjacent elements in the Periodic Table which have isotopes such that if the percentage isotopic compositions of the naturally occurring elements were different, their average atomic weights would stand in reversed order. Our present theories about the stability and origin of isotopes in the earth's crust have given a better explanation of Mendeleev's success, but even now the correspondence is still somewhat *ad hoc*.

Although the General Correspondence Principle provides a successful framework for describing structural relationships between successive mature theories it does not give us an account of the processes by which the theoretical change occurs. Also by focussing on S^* it tends to underemphasize the radical changes which do occur during scientific revolutions. Let us now turn to a more detailed description of the discontinuities in the history of science.

II. A DESCRIPTION OF THE ROLE OF PREFERRED STATEMENTS IN SCIENCE

Scientific revolutions, like political revolutions, only arise when a wholesale revision of the entrenched system is necessary. On a Popperian view of methodology one would expect a "continuing revolution" in science unless we were lucky enough to happen onto a theory which survived severe testing over long periods of time.[10]

I think everyone would agree that there have been long periods of theoretical stability in science. Newton's laws and the laws of conservation of matter and energy had reigns that any monarch would envy.

Moreover, when one looks at the history of science in detail it seems doubtful that the research was being conducted with the primary aim of putting the central tenets of these theories directly to the test. Quite the contrary. It often looks as if scientists were bending over backwards to protect the most fundamental postulates of the theory from revision.

Many philosophers have noted the existence of these "preferred statements", as I shall call them, and they have been the source of much philosophical puzzlement. They are like logical truths or other analytical truths, such as definitions, because their denial seems unthinkable (or at least *unacceptable*), but they are unlike traditional analytic statements because they are descriptive. In fact what they assert about the world is considered to be of central importance.

Hanson considered such statements to be conceptual truths. Feyerabend at one time seems to have thought of them as meaning postulates. Quine would place them near the relatively-analytic center of his net. Lakatos treats them as statements which have been rendered immune to refutation by a methodological decision – they are in the hard-core of a research program. They would constitute an important part of a Kuhnian paradigm. Philosophers of an earlier period might have described them as regulative principles.

Regardless of exactly which philosophical analysis we give them, it is clear that two theories which have contradictory (or incommensurable, if you prefer) sets of preferred statements differ in a deep and important way. People who hold different sets of preferred statements disagree about what they consider to be the most fundamental and important truths about the world and they would adopt different strategies in the face of certain experimental results. Thus many of the italicized statements in their text books differ and often the experiments they would carry out in the laboratory would be different. Small wonder that Kuhn described such scientists as "living in different worlds".

It is clear that in general many of the preferred statements of the old S theory do not belong to S^*, the portion of the old theory which can be made consistent with the later theory L. Examples include statements about the force of the vacuum or phlogiston, or the claim that atomic weights are fundamental constants of nature. Some of S's preferred statements may be in S^*; that is, they can be modified so that they are consistent with the new theory (e.g., "The mass of an object is independent of its velocity"). But since statements with preferred status are considered to be so central to an understanding of the universe that they are protected from empirical revision (for a time at least), any tampering with them must be considered to be a fairly drastic change. It would appear that a theory of the role of preferred statements in the development of science is needed to supplement the account of inter-theory relations given by the General Correspondence Principle.

A. *The Special Status of Preferred Statements*

As indicated above, some philosophers have thought that preferred statements are to be characterized in terms of their special logical status. However, I will propose that they are best viewed as having a peculiar *methodological* status. My view is suggested by Lakatos' approach although it differs from his in certain significant respects.

According to Lakatos' theory (1970), much of scientific progress comes through what he calls "progressive research programs". Every research program contains a "hard-core", a set of theoretical statements which have been rendered immune to falsification by a methodological decision. In case of any discrepancy with experimental results, we resolve to modify the auxiliary hypotheses (or even reinterpret the experiment), not tinker with the core of the theory. Each research program also contains a "positive heuristic", a plan for developing the theory, often by making more and more elaborate mathematical models of complex physical situations. A research program is progressive as long as it continues to lead to new discoveries.

In as much as one has *decided* to make the hard-core statements an irrefutable part of the system, one could view them as conventional truths. Lakatos never says so explicity, but it may be that he thinks of them as analytic statements. Here are my reasons for thinking so. In 'Proofs and Refutations' (1963–64) he introduced the notion of "methodological analyticity" to describe the logical status of certain statements in mathematics – he may have a similar view of hard-core statements in science. Also one should recall, as Professor Lakatos kindly pointed out to me in conversation, that since hard-core statements have been declared immune to refutation, there are no potential falsifiers which could actually lead to their rejection. Therefore, according to one interpretation of Popper's definition they have no empirical content.

However, I reject such a pragmatic interpretation of empirical content. I suggest we should view the logical status of preferred statements as follows: They are synthetic, empirical statements which are logically related to a set of potential falsifiers. However, we have decided (for reasons which will be discussed later) that under certain circumstances, whenever an inconsistency in our system develops, we will delay modifying the preferred statements as long as possible. (Later I will discuss those circumstances under which one does modify or drop them.) Thus they are somewhat similar to Popper's basic statements. They are not declared true by definition, but they are taken as true unless a special case is made against them. There is an important difference, however. If we assert that "Roger is a swan" is a basic statement, it is because (i) we have decided that being a swan is the sort of thing which one can "safely" attribute to objects like Roger and (ii) we have investigated Roger. Thus "Roger is not a swan" could also have been a basic statement.[11] This parallel does *not* hold for preferred statements. On my account, if one were to formalize a scientific theory, one would not put a logical necessity operator in front of the preferred statements. However, they would be given some sort of gold star to indicate that for a variety of epistemic, methodological, or heuristic reasons, these statements are preferred ones.

Given this characterization of preferred statements one might well ask whether such statements ever play an important role in the development of science. Do theories really have hard-cores of preferred statements? Certainly Newton's Laws were around for a long time, but perhaps that was just because it was a very successful theory which did not get refuted for a couple of hundred years. Did anyone really *decide* not to allow Newton's Laws to be refuted? I have some sympathy with such questions. I think most accounts of paradigms or research programs have tended to make science appear more monolithic than it really is. Nevertheless, I think historical studies do show that there were certainly *prima facie* refutations of Newton's Laws and that in attempting to account for these, physicists were very reluctant to question Newton's Laws. There were some attempts to tinker with the inverse square law by adding extra terms, but this was certainly not the standard or prevailing practice.[12]

But there is a much more typical situation which we can describe only by introducing some notion of preferred statements. As Feyerabend has emphasized, new theories which are born into a sea of anomalies must be protected from severe criticism, at least until we have a chance to work them out in detail. For example, Bohr's theory was incompatible with classical electromagnetism; Copernicus' heliocentric theory was inconsistent with observation reports concerning the immobility of the earth; Prout's hypothesis was refuted by the best measured values of the atomic weights. Yet on the face of it all of these theories contributed to the growth of science.

If we grant that preferred statements appear to have played an important role in the development of science, we are faced with the problem of trying to give guidelines for the rational utilization of preferred statements. Unless some argument can be given *for* the decision to protect certain statements from criticism (even if only temporarily), the whole process would seem to be dogmatic and anti-critical – or at the very least arbitrary. Did Copernicus and Kepler have good reasons for assuming that the earth moved around the sun or were they merely lucky "sleep walkers"?

III. THE RATIONAL UTILIZATION OF PREFERRED STATEMENTS

In the history of science there have undoubtedly been a vast variety of considerations which have entered into the decision to give certain statements preferred status, not all of them of equal merit. Stahl probably gave preferred status to the statement that all materials were composed of three principles simply because Becher and other predecessors had said so. However, the statement that phlogiston was lost in combustion gained its preferred designation because it was the keystone to a theory which successfully correlated a whole host of chemical reactions, including not only combustion and calcination, but also the reduction of metal ores, reactions with nitre, the laboratory preparation of sulfur, etc. An even stronger case could be made for declaring Newton's Laws to be preferred statements. They explained and correlated too many phenomena, they solved too many problems, they were too conceptually and mathematically simple to be given up lightly.

Sometimes a statement may be preferred because of its heuristic importance. Prout's hypothesis that all atomic weights were integers might have been somewhat useful for summarizing and correcting experimental data, but its chief interest lay in its implications for further theoretical developments – if it were true, one could then look for the fundamental building block of matter and investigate its properties. Often there are also metaphysical motives for giving theoretical statements a preferred status – theories which are deterministic, or imply a unity of matter, or postulate the primacy of free-willed individuals often seem especially attractive.[13]

Although historically there have been many reasons for the decision to protect certain statements from refutation, there seem to be only two major sorts of arguments for such a decision. First, one can argue that if they *were* true, these statements would lead to or constitute a very interesting theoretical development (a heuristic argument). Secondly, one notes that these statements have already given a demonstration of their explanatory power, and argues that there is some reason to believe that the statements are true, or have a high degree of verisimilitude (an empirical argument).

We will clearly distinguish between two sorts of rational decisions concerning preferred statements even though actual historical cases generally involve a confused mixture of both. First, one may decide to conduct one's research *as if* a certain statement were true. Secondly, one may decide that there are certain parts of a theory which we have good reason to believe *are* true or have a high truth content. The rational ways to approach these two decisions are different. And the epistemological status of the preferred statements so chosen is different. In the first case one is committed at least temporarily to some sort of instrumentalism; in the second case one is entitled to view them as realistic claims. Let us now discuss the two cases in detail.

A. *The Decision to Prefer Statements Because of Their Promise*

The sort of argument used in Pascal's wager is not a reasonable way to decide what to believe (as Pascal may have thought it to be), but it is a good way to decide which things to act *as if* you believe. The decision to start pursing a certain line of research may arise out of deliberation which can be described using a matrix such as the following (actual cases will be more complex) – the rows represent possible acts, columns are possible states of nature, and A, B, C and D are the corresponding outcomes:

	T has a high truth content	*T* has a low truth content
Choose to protect *T*	A. Work on a theory which may eventually be shown to have a high truth content.	B. Work on a theory which will eventually be discarded (cf. Type II error).
Choose not to protect *T*	C. Ignore a theory which has a high truth content (cf. Type I error).	D. Ignore a theory with a low truth content.

I have described the truth of the theory in question in terms of truth content instead of saying it is either true or false *simpliciter* for two reasons. First, it is generally the case that the only reasonable estimate of the probability of a particular theory's being literally true is zero, but that need not be the case of our judgments concerning the probability that it has a fairly high truth content. Secondly, the General Correspondence Principle reminds us that scientific progress is usually made via theories which are at best only approximately true.

I will not try to give a detailed theory about how one estimates the probabilities and utilities of these four outcomes. The methods used will depend on one's basic philosophical position. Neither am I interested here in expanding the matrix to include considerations such as the availability of alternative theories or the size and flexibility of the scientific community. (If one believes that other scientists will start up alternative research programs, the disutility of one scientist's going off down a blind alley will not be very large.)

Even this crude model illustrates several important points:

(1) There are cases in which it can be rational to work on a theory whose truth content, according to the best available evidence, is likely to be low – namely cases in which the scientific promise of the theory is very high. This is how I would account for Feyerabend's examples of the "counter-inductive" methods of scientists.

(2) Although one may rationally embark upon work on a theory which contradicts the available background knowledge, it is not the case that "anything goes". Unless a theory has a high degree of potential scientific merit, it will not be rational to work on it. This fact puts severe limitations on the Feyerabendian principle of proliferation.

(3) Although we may be realists in principle and hope that the research undertaken as a result of this sort of deliberation will give us descriptive theories, nevertheless, this type of argument leads us to act like instrumentalists. For example, it may very well happen that there are two inconsistent theories both of which have such potential merit that we decide to make each of them the basis of a research program. (Perhaps the research carried out on the wave and corpuscular theories of light in the 17th century is an example of this sort of situation.)

By introducing Pascal's wager into science I may have succeeded in explaining the rationality of scientists such as Copernicus or Prout who propose theories which are inconsistent with large quantities of accepted observational evidence, but only at the great cost of committing myself to a very unattractive form of instrumentalism! Must we conclude that although the aim of science is compatible with realism, the methods of science are instrumentalistic? Let us see.

B. *The Decision to Prefer Statements Because of Their Truth Content*

I have cited the cases of Copernicus and Prout as examples of decisions to pursue an exciting line of research in the face of strong counterevidence. Yet even in these cases it could be argued that from the outset there were fairly strong reasons to believe that the statements in question had a considerable truth content – Copernicus could give a very simple qualitative account of retrograde motion and Prout could argue that it was very improbable that on chance alone so many atomic wieghts would come so near to being integers.[14]

If the line of research undertaken in accordance with a decision based on both epistemological and heuristic considerations is successful, we eventually come to a situation in which it can reasonably be argued that the theory actually has a considerable degree of truth-content. After the theory has "proved" itself by correctly predicting surprising phenomena, solving outstanding scientific problems, giving a unified explanation of previously unconnected phenomena, etc., we still treat it as a preferred statement. That is, if a *prima facie* refutation comes up, we direct the *modus tollens* away from the hard-core of the theory just as we did when we first began to explore the potential of the theory. Now, however, our reasons for doing so are quite different. On epistemological grounds *alone* we decide that it is more likely that the mistake lies elsewhere.

In actual historical cases it may be somewhat artificial to demarcate the period in which statements are given preferred treatment because they are potential prodigies from the period in which the preferred status is based on past performance. Also the overt behavior of scientists in the two cases is similar. However, it is very important to keep the philosophical distinction quite clear. As we shall see later, this is one of the points at which Lakatos' theory of research programs goes astray.

IV. THE CRITICISM OF PREFERRED STATEMENTS

On our account so far, it would appear that the preferred statements of a successful theory are never subjected to severe criticism – the theory starts out as a sheltered prodigy and ends up being granted tenure.[15] How is it that preferred statements ever come to be rejected? Both Kuhn and Lakatos provide answers to this question.

Briefly, Kuhn's answer is that a crisis develops when there is a build-up of anomalies. If there are lots of puzzles which resist repeated attempted solutions, then we begin to question the wisdom of retaining the preferred statements of the paradigm.

Lakatos' answer is that refutations (regardless of how many of them there may be) play no role in the decision to stop protecting the hard-core: "Every theory is born into a sea of anomalies." *All* that is important is the number of new discoveries which a research program generates. It is only if confirmations stop coming in that we may consider jumping on a new band-wagon and changing our allegiance to a new research program.

I shall criticize and suggest modifications to the above views. The aim of this section is to provide an account of scientific revolutions which is both realistic and rational. I shall present three common methods of criticizing statements which have been given preferred status because of their achieved explanatory power.

A. *Falsification by Attrition*

According to Kuhn (1962), the accumulation of anomalies leads to a general dissatisfaction with the prevailing paradigm. However, this dissatisfaction does not arise because these wide-spread anomalies lead us to judge that the theory may be false – for Kuhn no question of truth arises. It is just that no one wants to work on a puzzle where it appears that some of the pieces are missing! However, one can interpret the Kuhn pattern in a different way. Let T be the preferred statements of a theory and suppose $T \& A_1$ imply R_1, but experiments indicate $\neg R_1$. In such a case, given that there are strong arguments for T, one may very well look for arguments against A_1 or $\neg R_1$. However, if it is discovered that

$$T \& A_2 \rightarrow R_2, \text{ but } \neg R_2,$$
$$\vdots$$
$$T \& A_n \rightarrow R_n, \text{ but } \neg R_n,$$

we may well judge it likely that T itself is false. The more varied the A's and the more independent support we have for them, the more reasons there are for blaming T. (For a more formal analysis of the logic of falsification by attrition, see Appendix I.)

I would suspect that the rejection of the phlogiston theory relied partly on falsification-by-attrition arguments. The discovery of each new common gas posed a problem for the phlogiston theory. To cite only one problem, fixed air (CO_2) was produced by the combustion of charcoal; hence it should be rich in phlogiston. Yet it could also be prepared from vitriolic acid and chalk, neither of which was supposed to be rich in phlogiston according to the theory.

Before proceeding to a description of other types of refutation of preferred statements let us look at Lakatos' view in more detail because he argues that negative appraisals play no role in the development of science. According to Lakatos, one should ignore any putative refutations of the theory as long as new confirmations keep coming in.

First, it should be noted that in Lakatos' own examples many of the striking confirmations which keep the research program in a progressive phase arise from solving the problems posed by *prima facie* refutations. The planet Neptune was only discovered because people were concerned to explain the discrepancies in the orbits of the known planets. But Lakatos does not deny that refutations may play a heuristic role in the development of a theory. What he does deny is that the persistence of anomalies (such as the problems with the moon's orbit) or the discovery of new anomalies (such as Mercury's perihelion) play any role whatsoever in the appraisal of research programs. The only factor, according to Lakatos, which can tell against a research program is the *lack* of successful new predictions. It is a rather asymmetric criterion – failure to explain away a refutation doesn't count against you, but success in accounting for a putative refutation counts for you![16]

Lakatos argues that this view gives a good description of paradigmatic scientific practice. He also maintains that as long as a theory leads to the discovery of new things about the world, it is not irrational to continue to use it. I find the second claim an amazing one – even an instrumentalist is concerned with the relative number of successes and failures.

However, Lakatos' theory has some limited plausibility if we consider the evaluation of research programs in their early stages of development. When we make a Pascal's wager type of decision to work with a theory we may very well point to any astonishing successes we have as evidence of its heuristic potential. But before we can assert the theory as a correct or nearly correct description of the world, we must try to estimate the strength of the arguments which can be made for its truth. For this judgment any counter-evidence becomes very important. Since every false synthetic statement has some truth content (i.e., there are some true synthetic statements which follow from it), any theory, no matter how ridiculous, may lead its adherents to new discoveries – astrologers, palmists, and readers of the Tarot continue to rack up some remarkable successes! But I take it that the aim of science is true theories, not just occasional amazing true predictions. So if there is good reason to believe that a mature theory is false, then it would seem to be both dogmatic and counter-productive to continue to hold its statements as preferred ones.

B. *Falsification by a Striking Counter-Instance*

So far I have mentioned only one line of argument which might be used to argue that the hard-core statements of a theory should not be given preferred status – this was based on an analysis of the Kuhn case of an accumulation of little anomalies in various areas – what I have called falsification by attrition.

But occasionally there are cases in which a single discovery tells so severely against a theory that it seems impossible to imagine that any tinkering with auxiliary hypotheses could save it. We might call such a process falsification by a striking counter-instance. In such cases the discrepancy between prediction and experiment is so gross that one can argue that no modification of auxiliary hypotheses could account for it. (See Appendix II for an analysis of this situation.)

It is difficult to find historical examples in which the rejection of a set of preferred statements is occasioned by a single set of experiments. Usually there exists a wide variety of criticisms of the theory. One example where a single criticism was very important is the case of Mendeleev's Periodic Law and the discovery of the rare earth elements.

According to the Periodic Law, if the elements were arranged in order of atomic weight, there would be a periodic repetition of certain chemical properties such that the elements could be arranged into families within a table. After the brilliant initial successes of predicting the existence of the three missing elements, chemists fell to the task of making more and more exact determinations of the atomic weights of elements that were slightly out of order. In many cases the values suggested by the Periodic Table were experimentally confirmed. However, there were three "reversed pairs" of elements where better quantitative analyses did not bring their atomic weights into line. In his textbook, Mendeleev discusses these cases in detail, but is obviously not too worried about them – the discrepancies were slight and the analytic data still open to suspicion. However, the discovery of the rare earths (eventually there were 14 of them) was a massive refutation of the Periodic Law. According to their atomic weights the rare earths should lie in a horizontal row across the Periodic Table, but because of their very similar chemical properties they should all lie in one family or column. As Mendeleev put it in the introduction to his book: "... this portion of the periodic system is, in a way, broken." (1905 – Vol. I, p. xvii). Elsewhere he writes:

I have not formed any precise opinion on this score, and this appears to me to be one of the most difficult problems offered to the periodic law. (1905 – Vol. II, p. 45)

In the text book, Mendeleev does not seem to have the heart even to discuss the rare earths himself. They are relegated to an appendix which was written by a Professor Brauner from Prague. Writing later, Soddy described the situation as

... a point-blank contradiction to the chemical principle of the Periodic Law ... in itself sufficient to contradict the original idea that it was the atomic weight that determined chemical character. (1954, p. 2)

Even after such a clear refutation, one continued to use the Periodic Law of course, reasoning that there must be some truth in it – all of its successes could not be just the result of chance. But one could not continue to consider it to be a law of nature. Might not other positions in the table also be occupied by a dozen similar elements instead of just one? Given the rare-earths precedent and in the absence of a better theory, no one could say. (For historical completeness one should add that there were additional criticisms of the Periodic Law, but they need not concern us here.)

I have suggested that a striking counter-instance may provide us with good reasons for criticizing preferred statements. However, Feyerabend has recently discussed historical cases where a theory has had what he calls "qualitative failures" and yet there was no thought of rejecting it (1970). One example he discusses is the failure of Newton's corpuscular theory of light to account for the fact that the rough surfaces of mirrors reflect light instead of scattering it. If Feyerabend (1970a) is right in saying that not even a plausible story to reconcile this inconsistency was forthcoming, then according to the theory I am developing here, there are only two possibilities. Either the corpuscular hypothesis should have been thought of as a somewhat promising instrument whose potential should be investigated or it should have been viewed as a realistic theory against which there was a sizeable or perhaps even critical amount of counter evidence. I have not had time to investigate this case in detail. Generally, of course, one neither simply accepts nor rejects a putative refutation but treats it as a problem which deserves further investigation, but the general principle is clear. There are only two sorts of good reasons for not accepting a putative refutation: (i) One can direct the *modus tollens* away from the core of the theory because one judges that the auxiliary claims are more apt to be false – in this case one is entitled to treat the theory as a realistic claim. (ii) One can direct the *modus tollens* away from the core of the theory because one is interested in developing the theory, finding out what novel predictions it makes, etc. – in this case one is entitled to view the theory as an interesting predicting instrument, but not as a proposed explanation.[17]

C. *Inter-Theoretic Criticism*

The two methods of criticizing preferred statements thus far discussed have both involved experimental results directly. One argues that there is good reason for taking the results of one type of experiment or a series of different experiments at face value and for considering the corresponding auxiliary assumptions to be true. Then preferred statements which are inconsistent with the conjunction of these are judged to be false.

There is a third variety of empirical criticism of preferred statements which is very powerful and which involves criticism via other theories. Consider a case in which some of the preferred statements within one theory T are inconsistent with the preferred statements of another theory T'. In such a case one cannot try to resolve the inconsistency within the total body of scientific knowledge by playing about with auxiliary hypotheses or repeating experiments with greater care. When there is a flat inconsistency between two sets of preferred statements at least one of the theories must be declared false. So inter-theoretic criticism is a very strong form of criticism indeed.

Such a situation may arise when we have two rival theories which give differing accounts of phenomena in largely overlapping domains. In such cases the decision need not be an agonizing one. One weighs the evidence for the two theories and judges which is the most likely to be true. Often there will be a correspondence relation between the two.

However, if the two inconsistent theories are in what we might call adjacent fields, such as chemistry and physics, the decision is a very difficult one. The following situation arose in the 1920's: One of the preferred statements in Bohr's theory said that electrons *moved* in closed orbits around the nucleus. According to Lewis' valence theory of chemical bonding, the electrons were located in a *static* geometric pattern around the nucleus and were shared by nuclei. Bohr's theory was having great success in accounting for spectra; Lewis' theory could account for many of the properties of chemical molecules, such as the bent configuration of the water molecule.[18]

In such a situation one is forced to conclude that at least one of the theories is false. Nevertheless, it may still be reasonable to believe that both theories have a high degree of verisimilitude. The GCP provides weak inductive support for this assumption in any case where each theory has received a high degree of corroboration before the severe difficulties set in.

Sometimes one can remove the inconsistency by qualifying one or both of the theories. And in particular cases there may even be information available as to the domain where and the degree to which the theory can be trusted. This information may come from experimental results or from other theories which place limitations on the validity of the theory in question. However, in all such cases it would be wrong to continue to describe the central tenets of the theory as having preferred status. If a strong case has been made that they are literally false, they must be used only as loose, approximate guides towards an understanding of the world until a better theory comes along to supersede and correct them. Often the two theories are replaced by an overarching theory which resolves the inconsistency and unifies the fields into a single coherent framework.

In this paper I have tried to describe typical features of developing science and to analyze the types of rational strategies that scientists employ in their research. Some readers will be disturbed at the claim (as indeed I am to some extent) that many of our most admired scientific theories make their debut in an instrumentalistic attire. On further reflection, however, this discovery is perhaps not so surprising. There are many occasions in science in which the field is in a turmoil and no one really wants to claim that a suitable explanatory theory is in hand, although there may be various likely candidates under investigation. One solution would be to relegate such situations to the context of discovery and say no more about them. Another solution is to claim that since scientists sometimes play around with inconsistent statements and *ad hoc* hypotheses, science is irrational. Whatever its shortcomings, I would suggest that the approach taken here is at least better than these.

Indiana University

APPENDIX I: THE LOGIC OF FALSIFICATION BY ATTRITITION

The problem is to characterize those circumstances in which it is reasonable to discredit the preferred statements of a theory (call them T) in the face of an experimental counter-example instead of directing the *modus tollens* to the auxiliary hypotheses (call them A).

Assume $T \cdot A \to e$.

A. Popper's formula for the explanatory power of h with respect to e is:

$$E(h, e, b) = \frac{p(e, h \cdot b)}{p(e, b)},$$

where b is background knowledge (1963, p. 391). (In one of the new appendices of *The Logic of Scientific Discovery*, Popper defines corroboration in terms of explanatory power [1934, pp. 400–402], but we will not need to use the more complicated formula here.) When this formula has been applied in the past, one has assumed that all auxiliary hypotheses were unproblematic and hence could be included in b. However, let us not make that assumption and develop the formula for the case where the A-component of the background knowledge *is* considered to be problematic.

Let us now substitute T for h, shorten the formula by dropping b from the explicit notation (every probability is assumed to be conditional on b), and include the possibility that A may be false by using the formula for total probability. We will also consider the situation relative to $\neg e$, since our primary concern is with what happens if the experiment does not agree with the prediction.

$$E(T, \neg e, A,) = \frac{p(\neg e, T \cdot A) \times p(A) + p(\neg e, T \cdot \neg A) \times p(\neg A)}{p(\neg e)}.$$

In our example $p(\neg e, T \cdot A) = 0$ since $T \cdot A \to e$. So,

$$E(T, \neg e, A,) = \frac{p(\neg e, T \cdot \neg A) \times p(\neg A)}{p(\neg e)}.$$

What conditions will tend to make this quotient very low, thereby lowering our estimate of the explanatory power of T?

(a) If $p(\neg e)$ is high. This is one of the standard Popperian requirements for a severe test.

(b) If $p(\neg A)$ is low. In the extreme case where A is in fact true, T would be completely falsified by the assertion of $\neg e$. If the fraction $p(\neg A)/p(\neg e)$ is very small, that is sufficient to insure the discorroboration of T.

Additional discorroboration of T will accrue if the predictions made with the help of additional well-corroborated auxiliary hypotheses are also refuted by experiment. (Cf. repeated application of Bayes' Formula.)

B. Let us now analyze the same situation using only the probability calculus and without relying on Popper's formula. Given that $T \cdot A \to e$, what *interesting* assumptions (of course, if one assumed $p(T) = 0$ and $p(A) \neq 0$, the result follows directly) must be made to insure that

(1) $\quad p(T \cdot \neg A, \neg e) < p(\neg T \cdot A, \neg e)$?

To phrase the problem another way, given the observation of $\neg e$, under what circumstances must the theory be discredited more than the auxiliary hypothesis? (I assume that we prefer not to abandon both I and A.)

First note that Formula (1) is equivalent to:

(2) $\quad \dfrac{p(T \cdot \neg A \cdot \neg e)}{p(\neg e)} < \dfrac{p(\neg T \cdot A \cdot \neg e)}{p(\neg e)}$

or, assuming $P(\neg e) \neq 0$

(3) $\quad p(T \cdot \neg A \cdot \neg e) < p(\neg T \cdot A \cdot \neg e)$.

But (3) can be rewritten as:

(4) $\quad p(\neg e, T \cdot \neg A) \times p(T \cdot \neg A) < p(\neg e, \neg T \cdot A) \times p(\neg T \cdot A)$

In order to insure that an inequality of the form

$$a \cdot b < c \cdot d$$

is true (assuming all the factors are positive), we only need insure that

$$a < c \quad \text{and} \quad b < d.$$

It is easy to show that $p(T \cdot \neg A) < p(\neg T \cdot A)$ if $p(T) < p(A)$. Assume:

(5) $\quad p(T) < p(A)$ (Requirement 1).

Then, by the formula for total probability:

$$p(T \cdot A) + p(T \cdot \neg A) < p(A \cdot T) + p(A \cdot \neg T);$$

hence,

(6) $\quad p(T \cdot \neg A) < p(\neg T \cdot A)$.

We must also require that the remaining two terms in inequality (4) are in a proper relationship, i.e., that:

(7) $\quad p(\neg e, T \cdot \neg A) < p(\neg e, \neg T \cdot A)$

or:

(8) $\quad p(e, T \cdot \neg A) > p(e, \neg T \cdot A)$ (Requirement 2).

Can we make the requirement given in (8) intuitively plausible? Recall that $p(e, T \cdot A) = 1$. Now Equation (8) seems to say that a prediction failure can only be blamed on T if T is more "important" in the derivation of e than A is, i.e., if A really is just an "auxiliary" which must be added to T to "complete" the derivation.

APPENDIX II: FALSIFICATION BY A STRIKING COUNTER-INSTANCE

How are we to understand cases in which it is claimed that a single counter-example is so "striking" that no amount of reasonable tinkering with auxiliary hypotheses could save the theory? The simple answer is when $p(A, b)$ is nearly 1. However, it is rare for the evidence for a particular set of auxiliary hypotheses to be that strong.

What often happens in the case of a massive refutation is that one can argue that there is a very large set of alternatives to A, (let $\mathscr{A}$ be the set: $A \vee A' \vee A'' \vee A''' \ldots$) such that none of the disjuncts of $\mathscr{A}$ will save T and such that $p(\mathscr{A}, b)$ is very high indeed. For example, if one were to discover a small satellite which moved around the moon in a square orbit, one might argue that there is no reasonable set of boundary conditions which would save Newton's theory (or Einstein's for that matter).

Such examples are rare in the history of science. Falsification by attrition is much more common.

NOTES

* An earlier version of this paper was read at the Annual Meeting of the British Society for the Philosophy of Science, September, 1971.
[1] See Lakatos' criticism (1971, p. 179) of my (1971) paper.
[2] Post (1971) cites several physicists to this effect. For a detailed analysis of the relations between various areas of contemporary physics, see Tisza (1963). Strauss (1968) proposes a general schema to describe inter-theory relations.
[3] A detailed analysis of this case is found in my dissertation (1969). Galileo's theory is presented in his *Dialogues Concerning Two New Sciences*. Pascal's early experiments are described in *Expériences Nouvelles Touchant le Vide*. *The Equilibrium of Liquids and the Weight of the Mass of the Air* is of great interest because in it Pascal sets out parallel explanations of various phenomena, one given in terms of the force of the vacuum, the other in terms of atmospheric pressure.

⁴ My discussion of the General Correspondence Principle relies heavily on Post (1971), as well as his lectures and tutorials which I attended when I was his student at Chelsea College (London). However, since my dissertation also deals with the GCP, the precise way in which I approach the subject does not always coincide exactly with his. The reader is advised to consult his paper for a more detailed exposition of his views.

⁵ There are many places where Popper adopts a position very similar to the General Correspondence Principle. In one section of *The Logic of Scientific Discovery* Popper argues that his non-inductive theory of scientific method can give an explanation of the inductive direction in which science grows:

> "The quasi-inductive process should be envisaged as follows. Theories of some level of universality are proposed, and deductively tested; after that, theories of a higher level of universality are proposed and in their turn tested with the help of those of the previous levels of universality, and so on." (1934, p. 276)

> "... a theory which has been well corroborated can only be superseded by one of a higher level of universality; that is, by a theory which is better testable and which, in addition, *contains* the old well corroborated theory – or at least a good approximation to it." (1934, p. 276)

See also "The Aim of Science" (1957) and "What is Dialectic?" (1963) It should also be noted that the motto on the page facing the title essay in the second edition of *Conjectures and Refutations* is a quotation from Einstein: "There could be no fairer destiny for any...theory than that it should point the way to a more comprehensive theory in which it lives on, as a limiting case."

⁶ Whewell relied on the consilience of inductions in his formulation of a position somewhat similar to the GCP:

> "No example can be pointed out, in the whole history of science, so far as I am aware, in which this Consilience of Inductions has given testimony in favour of an hypothesis afterwards discovered to be false;...when the hypothesis, of itself and without adjustment for the purpose, gives us the rule and reason of a class of facts not contemplated in its construction, we have a criterion of its reality, which has never yet been produced in favour of falsehood." (1847 – Vol. II, pp. 67–68)

Although the strong claim in the above quotation would not so indicate, Whewell realized that a theory supported by a consilience of inductions might turn out to need correcting (in fact, he discussed the force of the vacuum example); but he insisted that such theories were "not altogether baseless":

> "... if our scheme has so much truth in it as to conjoin what is really connected, we may afterwards duly correct or limit the mechanism of this connexion. If our hypothesis renders a reason for the agreement of cases really similar, we may afterwards find this reason to be false, but we shall be able to translate it into the language of truth." (1847 – Vol. II, p. 60)

⁷ See Hesse (1970).
⁸ See my dissertation (1969) for a full analysis of this case.
⁹ Fine (1971) argues that at present there is no adequate theory of meaning change. According to some of the proposed criteria, very minor scientific developments result in changes of meaning. On other proposals there are practically no conceptual changes

in the history of science. If an acceptable theory of meaning change is ever proposed the Q condition might include a description of any slight meaning changes which occurred in the transition from the S theory to the L theory.

[10] The phrase "continuing revolution" is taken from the title of Agassi (1968). However, in his (1967) "Science in Flux: Footnotes to Popper", Agassi addresses himself to the problem of explaining the rather high degree of theoretical conservatism which actually is found in science. He examines several possible causes of this stability, including human sluggishness, dogmatism, conservative science education, and the difficulty of criticizing metaphysical frameworks. At one point he remarks that "if it turns out that one may explain the existing stability of science only by additional hypotheses, one may well prefer to condemn the stability than to add the auxiliary hypotheses". (1967, p. 323) I agree with Agassi in thinking that one job of philosophy of science is to criticize actual scientific practice, not just describe it. Nevertheless, it may be that there are good reasons for the persistence of scientific theories. My strategy in this paper is to investigate the extent to which these complicated historical situations can be reduced to a more familiar problems-conjectures-criticisms format.

[11] In Popper's theory, it is not usually the case that the negation of a basic statement is basic. However, he also points out (1963, p. 386) that this can occur, e.g., "There is now a full grown Great Dane in my study".

[12] See Musgrave (1971) for a rather impressive list of attempts to modify Newton's Laws, however.

[13] The theory of science developed here does not sufficiently emphasize the importance of principles of continuity, simplicity and invariance in the growth of science. Are we to say that such principles are "merely metaphysical" as if that disposed of them? Or can we claim that such principles have been empirically confirmed because they have figured importantly in many successful theories? Are there good methodological reasons for adopting them? At this point I am not at all sure which approach I would wish to take on this issue.

[14] Mr. Russell Smith of Chelsea College (London) pointed out to me the extreme improbability that the observed distribution of atomic weights should arise by chance.

[15] The tenure metaphor is suggested by a remark by Shimony:

"[A few theories] (such as classical mechanics and Euclidean physical geometry) were displaced under such honorable circumstances that they may be regarded as promoted to the rank of 'theory emeritus'..." (1970, p. 94).

[16] Lakatos claims to be a verificationist (1970), but if is he is one, it a verificationist of a peculiar stripe because his theory has no concept of disconfirming instances. His logic of confirmation is like the thermometers used in hospitals – the temperature recorded can only rise, never fall! The only cause for concern is any tapering off of upward motion. Evidently he thinks science, as well as capitalism, can afford to ignore its internal contradictions and simply rely on the power of positive thinking!

[17] Only statements which are preferred because we believe them to be true can serve as premises in explanations. Theories which we believe only to have a high degree of verisimilitude (i.e., they are known to have some falsity content) can serve as reliable instruments for technology, but they cannot be used in explanations. Uninvestigated theories which we judge to have a low degree of verisimilitude given our scanty available evidence may serve as heuristic potential. See Popper (1963, p. 391) for a formal discussion of verisimilitude.

[18] My 1971 paper discusses this case in more detail.

BIBLIOGRAPHY

Agassi, J., 1966, 'Sensationalism', *Mind* **75**, 1–24.
Agassi, J., 1967, 'Science in Flux: Footnotes to Popper', in *Boston Studies in the Philosophy of Science*, Vol. III (ed. by R. S. Cohen and M. W. Wartofsky), D. Reidel, Dordrecht, pp. 293–330.
Agassi, J., 1968, *The Continuing Revolution: A History of Physics from the Greeks to Einstein*, History of Science Series (ed. by D. A. Greenberg), McGraw-Hill, New York.
Duhem, Pierre, 1941, *The Aim and Structure of Physical Theory* (transl. by P. P. Wiener from the 2nd French ed.), Princeton Univ. Press, Princeton, 1962.
Feyerabend, P. K., 1962, 'Explanation, Reduction and Empiricism' in *Minnesota Studies in the Philosophy of Science*, Vol. III (ed. by H. Feigl and G. Maxwell) Univ. of Minnesota Press, Minneapolis, pp. 28–97.
Feyerabend, P. K., 1970a, 'Classical Empiricism', in *The Methodological Heritage of Newton* (ed. by R. E. Butts and J. W. Davis), Univ. of Toronto Press, Toronto, pp. 150–70.
Feyerabend, P. K., 1970b, 'Against Method: Outline of an Anarchistic Theory of Knowledge', in *Analyses of Theories and Methods of Physics and Psychology* (ed. by M. Radner and S. Winokur), Vol. IV of *Minnesota Studies in the Philosophy of Science*, pp. 17–130.
Fine, A., 1971, 'Meaning and Approximation', Paper read at the 1971 Summer Meeting of the British Society for the Philosophy of Science. (Forthcoming in the *University of Pittsburgh Series in the Philosophy of Science*.)
Hesse, M., 1970, 'An Inductive Logic of Theories', in *Analyses of Theories and Methods of Physics and Psychology* (ed. by M. Radner and S. Winokur), Vol. IV of *Minnesota Studies in the Philosophy of Science*, pp. 164–80.
Kaufmann, W., 1963, *Faith of a Heretic*. Anchor Books, Doubleday, New York.
Koertge, N., 1969, 'A Study of Relations between Scientific Theories: A Test of the General Correspondence Principle', Unpublished Ph.D. dissertation, University of London.
Koertge, N., 1971, 'Inter-Theoretic Criticism and the Growth of Science', *Boston Studies in the Philosophy of Science*, Vol. VIII (ed. by R. C. Buck and R. S. Cohen), Reidel, Dordrecht, pp. 160–73.
Kuhn, T. S., 1962, *The Structure of Scientific Revolutions*, 1st ed., Univ. of Chicago Press, Chicago.
Lakatos, I., 1963–64, 'Proofs and Refutations', *Brit. J. Phil. Sci.* **14**, 1–25, 120–39, 221–43, 296–342.
Lakatos, I., 1970, 'Falsification and the Methodology of Scientific Research Programmes', in *Criticism and the Growth of Knowledge* (ed. by I. Lakatos and A. Musgrave), Cambridge Univ. Press, London, pp. 91–195.
Lakatos, I., 1971, 'Replies to Critics', *Boston Studies in the Philosophy of Science*, Vol. VIII (ed. by R. C. Buck and R. S. Cohen), Reidel, Dordrecht, pp. 174–82.
Mendeleev, D., 1905, *Principles of Chemistry*, Vols. I and II. 3rd English ed. from 7th Russian ed. (transl. by C. Kamensky), Longmans, Green & Co, London.
Musgrave, A. E., 1971, 'Falsification and Its Critics', Forthcoming in the *Proceedings of the IVth International Congress for Logic, Methodology and Philosophy of Science*.
Popper, K. R., 1934, *The Logic of Scientific Discovery* (transl. of German original with new appendices and footnotes), Hutchinson, London, 1959.

Popper, K. R., 1957, 'The Aim of Science', *Ratio (Oxford)* I, 24–35.
Popper, K. R., 1963, *Conjectures and Refutations*, Routledge and Kegan Paul, London.
Post, H. R., 1971, 'Correspondence, Invariance and Heuristics: In Praise of Conservative Induction', *Stud. Hist. Phil. Sci.* **2**, pp. 213–55.
Shimony, A., 1970, 'Scientific Inference', in *The Nature and Function of Scientific Theories* (ed. by R. G. Colodny), Univ. of Pittsburgh Press, Pittsburgh, pp. 79–172.
Soddy, F., 1954, *Isotopes*, Modern Science Memoir, No. 33. Murray, London.
Strauss, M., 1970, 'Intertheory Relations' in *Proceedings of the 1968 Salzburg Colloquium in Philosophy of Science*, D. Reidel, Dordrecht, pp. 220–84.
Tisza, L., 1963, 'The Conceptual Structure of Physics', *Reviews of Modern Physics* **35**, 151–85.
Whewell, W., 1847, *The Philosophy of the Inductive Sciences*, Vols. I and II, Facsimile of 2nd ed. Johnson Reprint Co., London, 1967.

HILARY PUTNAM

EXPLANATION AND REFERENCE

I. GENERAL SIGNIFICANCE OF THE TOPIC

In this paper I try to contrast Marxist (and more broadly realist) theories of meaning with what may be called 'idealist' theories of meaning. But a word of explanation is clearly in order.

There is no Marxist 'theory of meaning' as such, but there are a series of remarks on the correspondence between concepts and things, on concepts, and on the impossibility of a priori knowledge in the writings of Engels[1] and Lenin[1] which clearly bear on problems of meaning and reference, and which constitute the starting point for such a theory. In particular, there is a passage[2] in which Engels makes the point that a concept may contain elements which are not true of the things which correspond to that concept. Engels' example is the concept *fish*. A contemporary scientific characterization of fish would include, Engels says, such properties as life under water and breathing through gills; yet lungfish and other anomalous species which lack these properties are classified as fish for scientific purposes. And Engels argues, I think correctly, that to stick to the letter of the 'definition' in applying the concept *fish* would be bad science. In short, Engels contends that:

(1) Our scientific conception (I would say 'stereotype') of a fish includes the property 'breathing through gills', but

(2) 'All fish breath through gills' is not true! (and, *a fortiori*, not analytic).

I do not wish to ascribe to Engels an anachronistic sophistication about contemporary logical issues, but without doing this it is fair to say on the basis of this argument that Engels *rejects* the model according to which such a concept as *fish* provides anything like analytically necessary and sufficient conditions for membership in a natural kind. Two further points are of importance: (1) The fact that the concept "natural kind *all* of whose members live under water, breath through gills, etc." does not strictly fit the natural kind Fish does not mean that the concept does not *correspond* to the natural kind Fish. As Engels puts it, the concept is not exactly correct (as a description of the corresponding natural kind) but that does not make it a *fiction*. (2) The concept is continually changing as a result of the impact of scientific discoveries, but that does not mean that it ceases to correspond to the same natural kind (which is itself, of course, also changing). Again, without attributing to Engels a sophisticated theory of meaning and reference, it is fair, I think, to restate the essential gist of these two points in the following way: concepts which are not strictly true of anything may yet refer to something; and concepts in different theories may refer to the same thing. Of these two points, the second is obvious for most realists; with a few possible exceptions (e.g. Paul Feyerabend), realists have held that there are successive scientific theories about the *same* things: about heat, about electricity, about electrons, and so forth; and this involves treating such terms as 'electricity' as *transtheoretical* terms, as Dudley Shapere has called them,[3] i.e., as terms that have the same reference in different theories. The first point is more controversial; the idea that concepts provide necessary and sufficient conditions for class membership has often been attacked but, nonetheless, constantly reappears. Without it, however, the other point is moot. Bohr took it for granted that there are (at every time) numbers p and q such that the (one dimensional) position of a particle is q and the (one dimensional) momentum is p; if this was part of the meaning of 'particle' for Bohr, and in addition, "part of the meaning" means "necessary condition for membership in the extension of a term", then electrons are *not* particles in Bohr's sense, and, indeed, there are *no* particles "in Bohr's sense". (And no "electrons" in Bohr's sense of 'electron', etc.) In fact, none of the terms in Bohr's theory referred! It follows on this account that we cannot say that present electron theory is a better theory of the same particles that Bohr was referring to. I take it that this is the line of thinking that Paul Feyerabend represents. On an account like Engels', however, Bohr would have been referring to electrons when he used the word 'electron', notwithstanding the fact that some of his beliefs about electrons were mistaken, and *we* are referring to those same particles notwithstanding the fact that some of our beliefs – even beliefs included in our scientific "definition" of the term 'electron' – may very likely turn out to be equally mistaken. This seems right to me, and likewise Shapere's recent emphasis on the idea that such terms as 'electron' are *transtheoretical* seems to me right and important. The main technical contribution of this paper will be a sketch of a theory of meaning which supports Engels' and Shapere's insights.

An "idealist" theory of meaning, as I am using the term, might go like this (in its simplest form): the meaning of such a sentence as 'electrons exist' is a function of certain *predictions* that can be derived from it (in a pure idealist theory, these would have to be predictions about *sensations*); these predictions are clearly a function of the *theory* in which the sentence occurs; thus 'electrons exist' has no meaning apart from this, that or the other theory, and it has a different meaning in different theories.

The question of "reference" is a harder one for an idealist: the essence of idealism is to view scientific theories and concepts as instruments for predicting sensations and not as representatives of real things and magnitudes. But a sophisticated idealist is likely to say that the question of reference is "trivial",[4] if one has a scientific language L containing the term 'electron', then one can certainly construct a metalanguage ML over it *a là* Tarski, and define "reference" in such a way that "'electron' refers to electrons" is a trivial theorem. But if different scientific theories T_1 and T_2 are associated with different formal languages L_1 and L_2 (as they must be if the words have different meanings in T_1 and T_2), then they will be associated with different *meta*-languages ML_1 and ML_2. In ML_1 we can say "'electron' refers to electrons", meaning that 'electron' in the sense of T_1 refers to electrons *in the sense of T_1*, and in ML_2 we can say "'electron' refers to electrons" meaning that 'electron' in the sense of T_2 refers to electrons *in the sense of T_2*; but there is no ML in which we can even express the statement that "electron" refers to the same entities in T_1 and T_2 – or, at least, no prescription for constructing such an ML has been provided by Positivist philosophers of science. In short, just as the idealist regards 'electron' as *theory dependent*, so does he regard the semantical notions of reference and truth as theory dependent; just as the Marxist (and, more generally, the realist) regards 'electron' as *trans-theoretical*, so does he regard truth and reference as trans-theoretical.

II. THE MEANING OF PHYSICAL MAGNITUDE TERMS

A. *A Causal Account of Meaning*

My purpose here is to sketch an account of the meaning of physical magnitude terms (e.g. 'temperature', 'electrical charge'); not an account of meaning in general, although I will try to indicate similarities between what is said here about these terms and what Kripke has said about proper names and what I have said elsewhere about natural kind words. (Kripke's work has come to me second hand; even so, I owe him a large debt for suggesting the idea of causal chains as the mechanism of reference.)

On a traditional view, any term has an intension and an extension. "Knowing the meaning" is having knowledge of the intension; what it is to "know" an intension (construed, usually, as an abstract entity of some kind) is never explained. The extension of the color term 'red', for example, is the class of red things; the intension, according to Carnap, is the property Red. Carnap spoke of "grasping" the intension of terms; what it would be the "grasp" the property Red was never explained; probably Carnap would have equated it with knowing how to verify sentences of the form 'x is red', but this comes from his theory of knowledge, not his writings on semantics. In any case, understanding words is a matter of having knowledge. Full linguistic competence in connection with a word may require more knowledge than just the intension; for example, syntactical knowledge, knowledge of co-occurence regularities, etc.; but linguistic competence, like understanding, is a matter of *knowledge* – not necessarily explicit knowledge – knowledge in the wide sense, implicit as well as explicit, "knowing how" as well as 'knowing that", skills and abilities as well as facts, but all *knowledge* none the less.

According to the theory I shall present this is fundamentally wrong. Linguistic competence and understanding are not just *knowledge*. To have linguistic competence in connection with a term it is not sufficient, in general, to have the full battery of usual linguistic knowledge and skills; one must, in addition, be in the right sort of relationship to certain distinguished situations (normally, though not necessarily, situations in which the *referent* of the term is present). It is for this reason that this sort of theory is called a "causal theory" of meaning.

Coming to physical magnitude terms, what every user of the term 'electricity' knows is that electricity is a magnitude of some sort – and, in fact, not even that: electricity was thought at one time to possibly be a sort of substance, and so was heat. At any rate, speakers know that "electricity" and "heat" are putative physical *quantities* – capable of more and less, and capable of location. (I do not think that even these statements are *analytic*, but I think they have a kind of *linguistic* association with the terms in question.) In a developed semantic theory one might introduce a special semantic marker, e.g. 'physical quantity', for terms of this sort. I cannot, however, think of anything that *every* user of the term 'electricity' *has* to know except that electricity is (associated with the notion of being) a physical magnitude of some sort, and, possibly, that "electricity" (or electrical charge or charges) is capable of flow or motion. Benjamin Franklin knew that "electricity" was manifested in the form of sparks and lightning bolts; someone else might know about currents and electromagnets; someone else might know about atoms consisting of positively and negatively charged particles. They could all use the term 'electricity' without there being a discernible "intension" that they all share. I want to suggest that what they do have in common is this: that each of them is connected by a certain kind of causal chain to a situation in which a *description* of electricity is given, and generally a *causal* description – that is, one which singles out electricity as *the* physical magnitude *responsible* for certain effects in a certain way.

Thus, suppose I were standing next to Ben Franklin as he performed his famous experiment. Suppose he told me that "electricity" is a physical quantity which behaves in certain respects like a liquid (if he were a mathematician he might say "obeys an equation of continuity"); that it collects in clouds, and then, when a critical point of some kind is reached, a large quantity flows from the cloud to the earth in the form of a lightning bolt; that it runs along (or perhaps "through") his metal kite string; etc. He would have given me an *approximately correct definite description* of a physical magnitude. I could now use the term 'electricity' myself. Let us call this event – my acquiring the ability to use the term 'electricity' in this way – an *introducing event*. It is clear that each of my later uses will be causally connected to this introducing event, as long as those uses exemplify the ability I acquired in that introducing event. Even if I use the term so often that I forget when I first learned it, the intention to refer to the same magnitude that I referred to in the past by using the word links my present use to those earlier uses, and indeed the word's being in my present vocabulary at all is a causal product of earlier events – ultimately of the introducing event. If I teach the word to someone else by telling him that the word 'electricity' is the name of a physical magnitude, and by telling him certain facts about it which do not constitute a causal description – e.g., I might tell him that like charges repel and unlike charges attract, and that atoms consist of a nucleus with one kind of charge surrounded by satellite electrons with the opposite kind of charge – even if the facts I tell him do not constitute a definite description of any kind, let alone a causal description – still, the word's being in his vocabulary will be causally linked to its being in my vocabulary, and hence, ultimately, to an introducing event.

I said before that different speakers use the word 'electricity' without their being a discernible "intension" that they all share. If an "intension" is anything like a necessary and sufficient condition, then I think that this is right. But it does not follow that there are no ideas about electricity which are in some way linguistically associated with the word. Just as the idea that tigers are striped is linguistically associated with the word 'tiger', so it seems that some idea that "electricity" (i.e., electric charge or charges) is capable of flow or motion *is* linguistically associated with 'electricity'. And perhaps this is all – apart from being a physical magnitude or quantity in the sense described before – that is linguistically associated with the word.

Now then, if anyone knows that 'electricity' is the name of a physical quantity, and his use of the word is connected by the sort of causal chain I described before to an introducing event in which the causal description given was, in fact, a causal description of electricity, then we have a clear basis for saying that he uses the word to refer to electricity. Even if the causal description failed to describe electricity, if there is good reason to treat it as a mis-description *of electricity* (rather than as a description of nothing at all) – for example, if electricity was described as the physical magnitude with such-and-such properties which is responsible for such-and-such effects, where in fact electricity is not responsible for the effects in question, and the speaker intended to refer to the magnitude responsible for those effects, but mistakenly added the incorrect information "electricity has such-and-such properties" because he mistakenly thought that the magnitude responsible for those effects had those further properties – we still have a basis for saying that both the original speaker and the persons to whom he teaches the word use the word to refer to electricity.

If a number of speakers use the word 'electricity' to refer to electricity, and, in addition, they have the standard sorts of associations with the word – that it refers to a magnitude which can move or flow – then, I suggest, the question of whether it has "the same meaning" in their various idiolects simply does not arise. If a word is linguistically associated with a necessary and sufficient condition in the way that 'bachelor' is, then that sort of question *can* arise; but it does not arise, for example, in the case of proper names, and it does not arise, for a similar reason, in the case of physical magnitude terms. Thus if you know that 'Quine' is a name and I know that 'Quine' is a name and, in addition, we both refer to the same person when we use the word (even if the causal chains linking us to the referent are quite different) then the question of whether 'Quine' has the same meaning in my idiolect and in yours does not arise. More precisely: if the referent is the same, and we both associate the same minimal linguistic information with the word 'Quine', namely that it is a person's name, then the word is treated as the same word whether it occurs in your idiolect or in mine. Similarly, 'electricity' is the same word in Ben Franklin's idiolect and in mine. Of course, if you had wrong linguistic ideas about the name 'Quine' – for example, if you thought 'Quine' was a female name (not just that Quine was a woman, but that the name was restricted to females) – then there would be a difference in meaning.

This account stresses causal descriptions because physical magnitudes are invariably discovered through their effects, and so the natural way to first single out a physical magnitude is as the magnitude responsible for certain effects. Of course, the words 'responsible', 'causes', etc., do not literally have to occur in the description: *spin*, for example, was introduced by describing it as a physical magnitude having half-integral values characteristic of certain elementary particles, and giving a *law* connecting it with magnitudes previously introduced; I intend the notion of a causal description to include this case. And it is not a "necessary truth" that the description introducing a new physical magnitude should involve a notion of cause or law; but I am not trying in this paper to state "necessary truths".

Once the term 'electricity' has been introduced into someone's vocabulary (or into his "idiolect", as the dialect of a single speaker is called) whether by an introducing event, or by his learning the word from someone who learned it via an introducing event, or by his learning the word from someone linked by a chain of such transmissions to an introducing event, the referent in that person's idiolect is also fixed, even if no knowledge that that person has fixed it. And once the referent is fixed, one can use the word to formulate any number of theories about that referent (and even to formulate theoretical definitions of that referent which may be correct or incorrect scientific characterizations of that referent), without the word's being in any sense a different word in different theories. Thus the account just given fullfils the desideratum with which we started – it makes such terms as 'electricity' trans-theoretical. The "operational criteria" you can give for the presence of electricity will depend strongly on what theory you accept; but, without the illicit identification of meaning with operational criteria, it does not follow at all that *meaning* depends on the theory you accept.

The possibility of formulating definite descriptions (or even misdescriptions) of physical magnitudes depends upon the availability in our language of such "broad spectrum" notions as *physical magnitude* and *causes*; that these play a crucial role in the introduction of physical magnitude terms was argued in a previous paper.[5] In that paper, however, I did not distinguish between *defining* what I then called theoretical terms and *introducing* them. Of course, if we have available a language in which we can formulate descriptions of the referents of our various physical magnitude terms, then we can consider the various theories that we have containing those terms as so many different systems of sentences in that one language. To the extent that we can do this, we can treat the notions of reference and truth appropriate to that language as trans-theoretical notions also.

B. *Kripke's Theory of Proper Names*

I have already acknowledged a heavy indebtedness to Kripke's (unpublished) work on proper names. Since I have heard mainly secondhand reports of that work, I shall not attempt to describe it here in any great detail. But, as it has come down to me, the key idea is that a person may use a proper name to refer to a thing or person X even though he has *no* true beliefs about X. For example, suppose someone asks me who Quine is, and I falsely tell him that Quine was a Roman emperor. If he believes me, and if he goes on to use the word 'Quine' with the intention of referring to the person to whom *I* refer as Quine, then he will say such things as "Quine was a Roman emperor" – and he will be referring to a contemporary logician. Of course, he still has some true beliefs about Quine (beyond the belief that Quine is or was a person); for example, that Quine is or was named 'Quine'; but Kripke has more elaborate examples to show that even this is not always the case. On Kripke's view, the essential thing is this: that the use of a proper name to refer involves the existence of a causal chain of a certain kind connecting the user of the name (and the particular event of his using the name) to the bearer of the name.

Now then, I do not feel that one should be quite as liberal as Kripke is with respect to the causal chains one allows. I do not see much point, for example, in saying that someone is referring to Quine when he uses the name 'Quine' if he thinks that "Quine" was a Roman emperor, and that is all he "knows" about Quine; unless one has *some* beliefs about the bearer of the name which are true or approximately true, then it is at best idle to consider that the name refers to that bearer in one's idiolect. But what seems right about Kripke's account is that the knowledge an individual user of a language has need not at all fix the reference of the proper names in that individual's idiolect; the reference is fixed by the fact that that individual is causally linked to other individuals who were in a position to pick out the bearer of the name, or of some names from which the name descended. Indeed, what is important about Kripke's theory is not that the use of proper names is "causal" – what is not? – but that the use of proper names is *collective*. Anyone who uses a proper name to refer is, in a sense, a nember of a collective which had "contact" with the bearer of the name: If it is surprising that a particular member of the collective need not have had such contact, and need not even have any good idea of the bearer of the name, it is only surprising because we think of language as private property.

The relationship of this theory of Kripke's to the above theory of physical magnitude terms should be obvious. Indeed, one might say that physical magnitude terms *are* proper names: they are proper names of *magnitudes* not *things* – however, this would be wrong, I think, since some physical magnitude terms (e.g. 'heat') are linguistically associated with rather rich information about the referent. The inportant thing about proper names is that it would be ridiculous to think that having linguistic competence can be equated in their case with knowledge of a necessary and sufficient condition – thus one is led to search for something other than the knowledge of the speaker which fixes the referent in their case.

It will be noted that I required a causal chain from the use of the physical magnitude term back to an introducing event – not back to an event in which the physical magnitude played a significant role. The reason is that, although no one in practice is going to be in a position to give a definite description of a physical magnitude unless he is causally connected to such an event, the nature of *that* causal chain seems not to matter. As long as one is in a positon to give a definite description (or even a misdescription), one is in a position to introduce the term; and the chain from there on is something about which much more definite statements can be made. (In my opinion, it would be good to make a similar modification in Kripke's theory of proper names.)

C. *Natural Kind Words*

In an earlier paper[6] I presented an account of natural kind words (e.g. 'lemon') which has some relation to the present account of physical magnitude terms. I suggested that anyone who has linguistic competence in connection with 'lemon' satisfies three conditions: (1) He has implicit knowledge of such facts as the fact that 'lemon' is a concrete noun, that it is the "name of a fruit", etc. – information given by classifing the word under certain natural syntactic and semantic "markers". I criticized Jerrold Katz for the view that natural systems of semantic markers can enable us to give the exact meaning of each term (or of *any* natural kind term); but *some* of the information associated with a word can naturally be represented by classifying the word under such familiar headings as 'noun', 'concrete', etc. (2) He associates the word with a certain "stereotype" – yellow color, tart taste, thick peel, etc. (3) He uses the word to *refer* to a certain natural kind – say, a natural kind of fruit whose most essential feature, from a biologist's point of view, might be a certain kind of DNA.

Two points were most important in the argument of that paper. The first was that the properties mentioned in the stereotype (and, I would add, the properties indicated by the semantic markers) are not being analytically predicated of each member of the extension, or, indeed, of any members of the extension. It is not analytic that all tigers have stripes, nor that some tigers have stripes; it is not analytic that all lemons are yellow, nor that some lemons are yellow; it is not even analytic that tigers are animals or that lemons are fruits. The stereotype is *associated* with the word; it is not a necessary and sufficient condition for membership in the corresponding class, nor even for being a normal member of the corresponding class. Engels' example of the word 'fish' fits right in here: what Engels was pointing out was precisely that the stereotype associated with the term 'fish' even in scientific, as opposed to lay, usage is not a necessary and sufficient condition. The second point was that speakers must be referring to a particular natural kind for us to treat them as using the same word 'lemon', or 'aluminum', or whatever. The weakness of that paper, apart from being very poorly organized and presented, is that nothing positive is said about the conditions under which a speaker who uses a word (say 'aluminum' or 'elm tree') is referring to one set of things rather than another. Clearly, the speaker who uses the word 'aluminum' need not be able to tell aluminum from molybdenum, and the speaker who use the term 'elm tree' cannot tell elm trees from beech trees if he happens to be me. But then what does determine the reference of the terms 'aluminum', and 'molybdenum' in my idiolect? In the previous paper, I suggested that the reference is fixed by a test known to experts; it now seems to me that this is just a special case of my use being causally connected to an introducing event. For natural kind words too, then, linguistic competence is a matter of knowledge plus causal connection to introducing events (and ultimately to members of the natural kind itself). And this is so for the same reason as in the case of physical magnitude terms; namely, that the use of a natural kind word involves in many cases membership in a "collective" which has contact with the natural kind, which knows of tests for membership in the natural kind, etc., only as a collective. The idea that linguistic competence in connection with a natural kind word involves more than just having the right extension or reference (where this is now explained via a causal account), but also associating the right stereotype seems to me to carry over to physical magnitude words. Natural kind words can be associated with "strong" stereotypes (stereotypes that give a strong picture of a stereotypical member – even to the point of enabling one to tell, in most cases, if something belongs to the natural kind), as in the case of 'lemon' or 'tiger', or with "weak" stereotypes (stereotypes that give no idea of what a sufficient condition for membership in the class would be) as in the case of 'molybdenum' or (unless I am a very atypical speaker) 'elm'. Similarly, it seems to me that the physical magnitude term 'temperature' is associated with a very strong stereotype, and 'electricity' with a weak one.

D. *Objections and Questions*

It is obvious that the account presented here must face certain hard questions. Without attempting to think of all of them myself, I should like to list a few that may help to launch discussion.

(1) One question that must be faced by all causal theories of meaning is how to make more precise the notion of a causal chain of the appropriate kind. How precisely can we describe the sorts of causal chains that must exist from one use of a word to a later use of the same word if we are to say that the referent or referents are the same in the two cases? And how much of a defect in these sorts of theories is it if one cannot be more precise on this point?

(2) It may seem counterintuitive that a natural kind word such as 'horse' is sharply distinguished from a term for a fictitious or non-existent natural kind such as 'unicorn', and that a physical magnitude term such as 'electricity' is sharply distinguished from a term for a fictitious or non-existent physical magnitude or substance such as "phlogiston". Indeed, I myself believe that if unicorns were found to exist and people began to discover facts about them, give non-obvious definite descriptions or approximately correct descriptions of the class of unicorns, etc., then the linguistic character of the word unicorn would change; and similarly with 'phlogiston'; but this is certain to be controversial.

(3) Some people will argue that definitions of such terms as 'electricity' (or, more precisely, 'charge') are crucial in the exact sciences, and further that such definitions should be regarded as *meaning stipulations*. I agree with the first part of this – that definitions are important in science, provided one remembers what Quine has pointed out, that "definition" is relative to a particular text or presentation, and that there is no such thing, in general, as the definition of a term "in physics" or "in biology" – only the definition in X, Y, or Z's presentation or axiomatization. I disagree with the last part – that "definitions" in science are meaning stipulations – but, again, this is certain to be controversial.

(4) Finally, there will be objections to my use of causal notions, from Humeans who expect them to be reduced away, and to my use of the term 'physical magnitude' from extensionalists and nominalists. Here I can only plead guilty to the belief that talk about what causes what, or what the laws of nature are, or what would happen if other things hap-

pened is *not* highly derived talk about mere regularities, and to the belief that the real world requires for its description not only reference to things but reference to physical magnitudes [7] – in a sense of 'physical magnitude' in which physical magnitudes exist contingently, not as a matter of logical necessity, and in which magnitudes can be synthetically identical (e.g., temperature is the same magnitude as mean molecular kinetic energy).

III. WHY POSITIVISTIC THEORY OF SCIENCE IS WRONG

My contention in this paper is not that what is wrong with positivist theory of science is positivist theory of meaning. What is wrong with positivist theory of science today, as it was wrong with the Machian theories that Lenin criticized in 1908, [8] is that it is based on an idealist or idealist-tending world view, and that that world view does not correspond to reality. However, the idealist element in contemporary positivism enters precisely through the theory of meaning; thus part of any realist critique of positivism has to include at least a sketch of rival theory. In the present section, I want to turn from the task of sketching such a rival theory, which was just completed, to the task of showing that positivistic theory of explanation broadly construed – that is, positivist theory of scientific theory – does not correspond to reality any better than the older and less sophisticated idealist theories to which it is historically the successor.

Let us for a moment review some of those older theories. The oldest theory is Bishop Berkeley's. Here one already meets what might be called the *adequacy claim*: that is, the claim that a convinced Berkelian is *entitled* to accept standard scientific theory and practice, that Berkeley can give an account of the scientific method which would justify this. Indeed, I have heard philosophers argue that acceptance of Berkeley's metaphysics would not make any difference to the scientific theories one would accept. Here one already meets an important ambiguity. One can be claiming that a Berkelian can make the move of "accepting" scientific theory in some sense other than accepting as true or approximately-true: say, accepting as a useful prediction heuristic. If this is what one means, then the claim is trivial. To be sure, Berkeley can "accept" Newtonian physics in the Pickwickian sense of "accept" as a useful scheme for making predictions. But Berkeley, to do him justice, was interested in much more: What he claimed was that an idealist could *reinterpret* (only he would not consider it *re*-interpretation, but rather *correct* interpretation) the notion of object so as to square both the layman's and the scientist's talk of objects with the idealist claim that reality consists of minds and their sensations ("spirits" and their "ideas"). The difference between the two claims is the difference between accepting the idea that social practice is the test of truth and rejecting it, between accepting the idea that the overwhelming success of scientific theory offers some reason for accepting that theory as true or approximately-true, and claiming that success in practice is *no* indication of truth. Machian positivism fails for the same reason that Berkelian idealism does: although Mach makes the claim that his construction of the world out of sensations ("Empfindungen") is compatible with lay and scientific object-talk, no demonstration at all is given that this is so. The first philosopher to both precisely state and to undertake the task of *translating* thing-language into phenomenalistic language was Carnap (in *Logische Aufbau der Welt*). And what does Carnap do? He devotes the entire book to *preliminaries*, to "reconstructions" *within* sensationalistic language (i.e. reductions of some sensation-concepts to others, not of thing-concepts to sensation-concepts), and then in the last chapter gives a sketch of the relation of thing-language to sensation-language which is *not* a translation, and which, indeed, amounts to no more than the old claim that we pick the thing-theory that is "simplest" and most useful. In short, no demonstration is given at all that the positivist is entitled to quantify over (or refer to) material things.

It is with the failure of the phenomenalist translation enterprise, that is, with the failure to find *any* interpretation of object-concepts under which the prima facie incompatibility between an idealist world-view and a materialist world-view, between a world consisting of "spirits and their ideas", or of "Empfindungen", or of total experience-slices in one "specious present", and a world consisting of fields and particles, simply *disappears* – it is with this failure that contemporary positivistic philosophy of science begins. Basically, two moves were made by the positivists after the failure of phenomenalist translation. The first was to give up construing scientific theories as systems of statements each of which had to have an intelligible interpretation (intelligible from the standpoint of what was taken as "completely understood" or "fully interpreted"), and to construe them rather as mere calculi, whose objective was to give successful predictions and otherwise to be as "simple" as possible. "Scientific theories are partially interpreted calculi". [9] The second move was to shift from phenomenalist language to "observable thing language" as one's reduction-base – i.e., to say that one was seeking an interpretation or "partial interpretation" of physical theory in "observable thing language", not in "sensationalistic language".

The second move may make it appear questionable whether positivism is still correctly characterized as an "idealist" tendency – i.e., as a tendency which regards or tends to regard the "hard facts" as just facts about actual and potential *experiences*, and all other talk as somehow just highly derived talk about actual and potential experiences. I, myself, think this characterization *is* still fundamentally correct despite the shift to "observable thing predicates" for two reasons: (1) The cut between observable things and "theoretical entities" was historically introduced as a substitute for the thing/sensation dichotomy. Indeed, the reduction of "theoretical entities" to "observable things and qualities" would hardly seem to be a natural problem to someone who did not have in the back of his head the older problem of reduction to *sensations*. The reduction of things to sensations is both a historically motivated problem and one which rests upon the sharpness of the distinction between a material thing and a sensation (of course, even this sharpness is partly an illusion, on a materialist view – substitute "material process" for material thing!), as well as the supposed "certainty" one has concerning one's own sensations. But the reduction of electrons to tables and chairs, or, more generally, of "unobservable" things to "observable" things is not historically motivated, the distinction is not sharp (Grover Maxwell asked years ago if a dust mote is something "given" when it is just big enough to see and a "construct" when it is just too small to see – can the distinction between data and constructs be a matter of size?), and one is not supposed to have certainty concerning observable things. (2) The positivists themselves frequently say that one could carry their analysis back down to the level of sensations, and that stopping with "observable thing predicates" is a matter of *convenience*.[10]

In the remainder of this section I want to show that the first move – construing scientific theories as partially interpreted calculi – does not solve the adequacy problem at all. The positivist today is no more entitled than Berkeley was to accept scientific theory and practice – that is, his own story leads to no reason to think either that scientific theory is true, or that scientific practice tends to discover truth. In a sense, this is immediate. The positivist does not claim that scientific theory is "true" in any trans-theoretic sense of "true"; the only trans-theoretic notions he has are of the order of "leads to successful prediction" and "is simple". Like the Berkelian, he has to fall back on the position that scientific theory is *useful* rather than true or approximately-true. But he does try to provide some account of the acceptability of scientific theories, even some account of their "interpretation". And he wants to maintain that in some sense the principle on which Marx says realist philosophy of science rests – that social practice is the test of truth, that the success of scientific theories is reason to think they are true or approximately-true – is right. What I want to show is that the notion of "truth" that the positivist can give us is not the one on which scientific practice is based.

A. *Truth.*

When a realistically minded scientist – that is to say, a scientist *whose practice* is realistic, not one whose official "philosophy of science" is realistic – accepts a theory, he accepts it as true (or probably true, or approximately-true, or probably approximately-true). Since he also accepts *logic*[11] he knows that certain moves *preserve truth*. For example, if he accepts a theory T_1 as true and he accepts a theory T_2 as true, then he knows that T_1 & T_2 – the *conjunction* of T_1 and T_2 – is also true, by logic, and so he accepts T_1 & T_2. If we talk about probability, we have to say that if T_1 is very highly, probably true and T_2 is very highly probably true, then the conjunction T_1 & T_2 is also highly probable (though not *as* highly as the conjuncts separately), provided that T_1 is not negatively relevant to T_2 – i.e., provided that T_2 is not only highly probable on the evidence, but also no less probable on the added assumption of T_1 (this is a judgement that must be made on the basis of what T_1 *says* and of background knowledge, of course). If we talk about approximate-truth, then we have to say that the approximations probably involved in T_1 and T_2 need to be compatible for us to pass from the approximate-truth of T_1 and T_2 to the approximate-truth of their conjunction. None of these matters is at all deep, from a realist point of view. But even if we confine ourselves to the simplest case, the case in which we can neglect the chances of error and the presence of approximations, and treat the acceptance of T_1 and T_2 as simply the acceptance of them as true, I want to suggest that the move from this acceptance to the acceptance of the conjunction is one to which one is not entitled on positivist philosophy of science. One of the simplest moves that scientists daily make, a move they make as a matter of propositional logic, a move which is central if scientific inquiry is to have any *cumulative* character at all, is totally *arbitrary* if positivist philosophy of science is right.

The difficulty is very simple. Acceptance of T_1, for a positivist, means acceptance of the calculus T_1 as leading to succesful predictions (i.e., all *observation sentences* which are theorems of T_1 are true; not all *sentences* which are theorems of T_1 are "true" in any fixed trans-theoretic sense). Similarly, the acceptance of T_2 means the acceptance of T_2 as leading to successful predictions. But from the fact that T_1 leads to successful predictions and the fact that T_2 leads to successful predictions it does not follow at all that the conjunction T_1 & T_2 leads to successful predictions. The difficulty, in a nutshell, is that the predicate which plays the *role* of truth – the predicate "leads to successful predictions" – does not have the *properties* of truth. The positivist may teach in his philosophy seminar that acceptance of a scientific theory is acceptance of it as "simple and leading to true predictions", and then go out and do science (or his students may go out and do science) by verifying theories T_1 and T_2, conjoining theories which have been previously verified, etc. – but then there is just as great a discrepancy between what he teaches in his philosophy seminar and his *practice* as there was between Berkely's teaching that the world consisted of spirits and their ideas and continuing in practice to daily rely on the material object conceptual system.

Nor does it help to bring in "simplicity". It is not obvious that the conjunction of simple theories is simple; and even if simplicity is preserved by conjunction, the conjunction of simple theories which separately lead to no false predictions may even be *inconsistent* (examples are easy to construct). More sophisticated moves have indeed been made. Thus, for Carnap truth of a theory is the same as truth of its "Ramsey sentence"[12]. But exactly the same objection applies: "truth of the Ramsey sentence" does not have the properties of truth: if T_1 has a true Ramsey sentence and T_2 has a true Ramsey sentence it does not at all follow that the conjunction does.

(For those readers familiar with Carnaps' use of the Hilbert epsilon-symbol, it may be pointed out that the difficulty comes out in very sharp form in Carnap's symbolization of his interpretation of individual theoretical terms. Thus let $T_1(P)$, $T_2(P)$ be two theories containing exactly one theoretical term P. On Carnap's own symbolization of his view[12], what P means in T_1 is $\varepsilon PT_1(P)$; what P means in T_2 is $\varepsilon PT_2(P)$; and what P means in T_1 & T_2 is $\varepsilon P[T_1(P)$ & $T_2(P)]$; this makes it explicit that P has different meanings in T_1 and T_2 and *yet a third meaning* in their conjunction.)

B. *Simplicity.*

It is easy to construct a "theory" in the positivist sense (a calculus containing some observation terms) which leads to no false predictions but which no scientist would dream of accepting. This is usually handled by saying that scientists only choose "simple" theories. Also, a simple theory may mess up science as a whole: So it is said that scientists are trying to maximize the simplicity of "total science". "Theory" means, then, "formalization of total science, or of some piece which is independent of the rest of total science". Unfortunately, no one has ever written down or ever will write down a "theory" in this sense. The fact is, that positivist philosophy of science depends on a constant slide between giving the impression that one is talking about "theories" in the customary sense – Newton's theory, Maxwell's theory, Darwin's theory, Mendel's theory – and saying, at key points of difficulty such as the one just alluded to, that one is *really* talking about a "formalization of total science", or some such thing.

The difficulty with the rule "choose the simplest theory compatible with the evidence" is that it is probably not *right*, or would probably not be right, even if one *could* formalize "total science" (at a given time). Scientists are not trying to maximize some formal property of "simplicity"; they are trying to maximize *truth* (or improve their approximation to truth, or increase the amount of approximate-truth they know without decreasing the goodness of the approximation, and so forth).

Of course, a realist might accept the rule "chose the simplest hypothesis", if it could be shown that the simplest hypothesis is always the most *probable* on the basis of the rest of his knowledge. But this is not so on any usual measure of simplicity. For example, suppose I know just three points on interstate highway 40, and those three points lie on a straight line. Suppose also that the statement 'IS 40 is straight' is logically consistent with my total knowledge. Then accepting 'IS 40 is straight' would, on the usual simplicity metrics, be accepting the simplest hypothesis. Yet I would not in fact accept 'IS 40 straight', nor would anyone with our background knowledge. Given that every other interstate highway has curves, and given the enormous length of IS 40 and the enormous impracticality of making a straight highway across the entire U.S., it is overwhelmingly probable that IS 40 is *not* straight.

Can we not say that my *total* "knowledge" is less simple if I accept 'IS 40 is straight'? Not, it seems to me, on the basis of any criterion of *simplicity* that I know of. What is obviously involved here is not *simplicity* but plausibility: What introducing the word "simplicity" does is make it look as if a calculation which is in fact the calculation of the probability of a state of affairs is in reality just a calculation of a formal property (such as number of argument places, number of primitive symbols, length and number of the axioms, perhaps shape of the curves mentioned) of an uninterpreted or semi-interpreted *calculus*. Even if the property of being the most probable hypothesis on background knowledge could be *represented* syntactically, omitting to mention that the representing property was the syntactic representation of a *probability measure*, and pretending that it was *just* a formal property (like having simple axioms), would be a way of disguising rather than revealing what was going on.

C. *Confirmation.*

Indeed, positivist philosophers of science have made attempts at formalizing the logic of confirmation. These attempts are interesting (though so far unsuccessful) researches on *any* philosophy of science. But not only do they have nothing to do with positivist theory of meaning; they are in fact *incompatible* with it. Thus when they write about meaning, positivists tell us that "theoretical terms" have different meanings in different theories; when they formalize confirmation theory, they invariably treat theories as systems of sentences in *one* language, and assume that all semantical concepts are *trans*-theoretic. Thus the positivists are engaged in formalizing *realistic* confirmation theory; not the confirmation theory (if there is one!) to which their own theory of meaning should lead.

What is going on here should be evident from Carnap's work on the foundations of mathematics. Carnap has a consistent tendency to *identify* concepts with their syntactic representations: thus, mathematical truth with theoremhood (after the discovery of Gödel's theorem, he either allowed "non-constructive rules of proof", or simply assumed set theory, and took "logical consequence" rather than derivability as the basic notion, although this trivialized the "analysis" of mathematical truth). In the same way he would have liked to identify a state of affairs having a probability of, say, .9, with the corresponding sentences having a c-value of .9 (where "c" would be a syntactically defined measure on sentences in a formalized language). Even if Carnap had found a successful "c-function", the fact is that it would have been successful because it corresponded to a reasonable probability measure over some collection of states of affairs; but this is just what Carnap's positivism did not allow him to say.

D. *Auxiliary Hypotheses*

Sometimes, as we mentioned, the positivists make it explicit that the "theories" to which their theory of science applies are "formalizations of total science", and not theories in the usual sense; but their readers do, I think, tend to come away with the impression that their model *is* a model of a scientific theory in the usual sense – especially, a physical theory. Believing this involves believing that a physical theory is a calculus, or could easily be formalized as a calculus, and that its predictions are *self-contained* – that they are deduced from the explicitly stated assumptions of the theory itself. This leads to a comparison with social sciences which is derogatory to the social sciences – for the classic social science theories are clearly *not* self-contained in this sense. For example, when Marxists write that the capitalist class controls the state – that the army and the police intervene on the side of capitalists, that the politicians who have a chance of obtaining state power through elections are tied to capitalists, etc. – what they do is make a series of generalizations about ways in which this control allegedly takes place: what happens in elections, when the army intervenes, etc. But these generalizations do not lead to predictions about specific political events (or, not to predictions in the positivist's "observation language"), without very substantive auxiliary assumptions. These auxiliary assumptions, if they could be spelled out, would be numerous and would be highly context dependent. Thus Marxist theory must either be ignored or treated as not a scientific theory at all, but only a theory sketch, and likewise for such classic theories as those of Weber or even Mill. In short, the positivist attitude tends to be that social science is science only when and to the extent that it apes *physics*. And this for the reason that the mathematical model of a scientific theory provided by the positivists is thought to clearly fit *physical* theories.

But, in fact, it fits physical theories very badly, and this for the reason that even physical theories in the usual sense – e.g., Newton's Theory of Universal Gravitation, Maxwell's theory – lead to no predictions at all without a host of auxiliary assumptions, and moreover without auxiliary assumptions that are not at all law-like, but that are, in fact, assumptions about boundary conditions and initial conditions in the case of particular systems. Thus, if the claim that the term 'gravitation', for example, had a meaning which depended on the theory were true, and the theory included such auxiliary assumptions as that "space is a hard vacuum", and "there is no tenth planet in the solar system", then it would follow that discovery that space is *not* a hard vacuum or even that there is a tenth planet would change the meaning of 'gravitation'. I think one has to be pretty idealistic in one's intuitions to find this at all plausible! It is not so implausible that knowledge of the meaning of the term 'gravitation' involves some knowledge of the theory (although I think that this is wrong: the stereotype associated with 'gravitation' is not nearly as strong as a particular theory of gravitation), and this is probably what most readers think of when they encounter the claim that physical magnitude terms (usually called "theoretical terms" to prejudge just the issue this paper discusses) are "theory loaded"; but the actual meaning-dependence required by positivist meaning theory would be a dependence not just on the *laws* of the theory, but on the particular auxiliary assumptions – for, if these are not counted as part of the theory, then the whole theory-prediction scheme collapses at the outset.

Finally, neglect of the role that auxiliary assumptions actually play in science leads to a wholly incorrect idea of how a scientific theory is confirmed. Newton's theory of gravitation was not confirmed by checking predictions derived from it **plus** some set of auxiliary statements fixed in advance; rather the auxiliary assumptions had to be continually modified and expanded in the history of Celestial Mechanics. That scientific problems as often have the form of finding auxiliary hypotheses as they do of finding and checking predictions is something that has been too much neglected in philosophy of science;[13] this neglect is largely the result of the acceptance of the positivist model and its uncritical application to actual physical theories.

Harvard University

NOTES

[1] Cf. Lenin (1970) and Engels (1959).
[2] In a letter written to Conrad Schmidt in 1895; cf. Marx and Engels (1942), pp. 527–30.
[3] Cf. Shapere (1969).
[4] See, for example, the discussion by Hempel (1965), pp. 217–18. A contrasting view is sketched in Putnam (1962).
[5] Putnam (1962).
[6] Putnam (1970a).
[7] Cf. Putnam (1970b).
[8] Cf. Lenin (1970).
[9] Putnam (1962).
[10] E.g., Carnap says this on p. 63 in Carnap (1956).
[11] The role of logic in empirical science is discussed in Putnam (1971) and Putnam (1969).
[12] For details see Hempel (1965).
[13] I discuss this in 'The Corroboration of Theories', to appear in the forthcoming Popper volume in the *Library of Living Philosophers* series.

BIBLIOGRAPHY

Carnap, R., 1956, 'The Methodological Character of Theoretical Concept's, in *Minnesota Studies in the Philosophy of Science*, Vol. I (ed. by H. Feigl), Minneapolis.
Engels, F., 1959, *Anti-Dühring, Herr Eugen Dühring's Revolution in Science*, New-York.
Hempel, C., 1965, 'The Theoretician's Dilemma', in *Aspects of Scientific Explanation*, New York.
Lenin, V., 1970, *Materialism and Emperio-Criticism*, New York.
Marx, K. and Engels, F., *Selected Correspondence 1846–1895*, New York.
Putnam, H., 1962, 'What Theories Are Not', in *Logic, Methodology and Philosophy of Science* (ed. by E. Nagel, P. Suppes, and A. Tarski), Stanford.
Putnam, H., 1969, 'Is Logic Empirical?', in *Boston Studies in the Philosophy of Science*, Vol. V (ed. by R. Cohen and M. Wartofsky), New York.
Putnam, H., 1970a, 'Is Semantics Possible?', in *Language, Belief and Metaphysics*, Vol. I of *Contemporary Philosophical Thought: The International Philosophy Year Conferences at Brockport*, Albany.

Putnam, H., 1970b, 'On Properties', in *Essays in Honor of Carl G. Hempel* (ed. by N. Rescher), D. Reidel, Dordrecht, pp. 235–254.
Putnam, H., 1971, *Philosophy of Logic*, New York.
Shapere, D., 1969, 'Towards a Post-Positivistic Interpretation of Science', in *The Legacy of Logical Positivism* (ed. by P. Achenstein and S. Barker), Baltimore.

ROBERT BARRETT

REFERENTIAL INDETERMINACY: A RESPONSE TO PROFESSOR PUTNAM

Professor Putnam's paper is divided into two parts, in one of which (the latter, as it happens) he attacks what he identifies as Positivistic Theory of Science, and in the other of which (the first half) he sketches elements of an alternative theory of his own. I shall comment only on Putnam's theory, ignoring his attack on positivism. In thus taking advantage of the fact that philosophical positions tend to be relatively hard to defend and relatively easy to attack, I shall be attempting to make things relatively easy for myself and hard for Professor Putnam. First, let me outline briefly what I take to be the main tenets of Putnam's position.

He distinguishes "realist" from "idealist" theories of meaning, identifying a theory as idealistic when the concepts expressed by its terms are taken to play the semantically central role, determining, among other things, the referents of those terms; and as realistic when the referents of its terms are taken to play the semantically central role and to be independent of the concepts those terms express. His own theory of meaning is of the realistic variety.

In the company of Dudley Shapere and Friedrich Engels[1] Putnam wants to maintain that terms can and often do retain the same referents despite radical and pervasive changes in what other philosophers might call their "intensions". In particular, he wants to hold this for "physical magnitude" terms, that is, terms which, like 'electricity', 'temperature' and 'kinetic energy', may be said to refer to physical magnitudes. He wants to be able to say that such terms can and often do retain the same reference in spite of radical and pervasive changes in the scientific theory to which they belong, that 'electricity' continues to refer to just the same entity or entities throughout changes from the Rutherford to the Bohr theory of the atom, and that 'kinetic energy' has the same referent in relativistic mechanics that it has in classical Newtonian mechanics.

To accomplish this result, Putnam proposes a causal theory or reference for physical magnitude terms. According to this theory it is not concepts or intensions that determine the referents of these terms, but rather certain

Pearce and Maynard (eds.), Conceptual Change, 222–232. All rights reserved.
Copyright © 1973 by D. Reidel Publishing Company, Dordrecht-Holland.

events involving the referents themselves. These events are causally responsible for linguistic behavior in which the terms are employed. Or, if the word 'responsible' constitutes this too strong a claim, at least causal chains lead from these events directly to any given linguistic event in which such terms are used. Roughly speaking, the reference of a given physical magnitude term is determined by the very entity that is the referent of that term, through a causal chain leading from the latter to uses of the former.

Although I have given a very sketchy and incomplete account of Putnam's thesis, I think it will serve my purposes well enough. Basically, what I want to do is to challenge a single fundamental assumption on which the whole proposal appears to rest. Not that I do not share with Professor Putnam at least some rather broadly defined goals. I too should like to be able to regard alternative scientific theories as comparable, rather than in some sense incommensurate, and should hope to be able to provide some clear basis for their comparison. However, I do not think that this can be accomplished by appeal to the sameness of reference of key theoretical terms. This is not to say that I think it false that certain terms in a scientific theory retain the same reference in spite of radical and pervasive changes in the theory, nor that I think it false that the term 'electricity' continues to refer to just the same entity or entities throughout changes from the Rutherford to the Bohr theory of the atom. And it is not to say that I think it false that the term 'kinetic energy' has the same extension in relativistic mechanics that it has in classical Newtonian mechanics.

Rather, I find each of these claims neither true nor false. I have no inclination to paraphrase this remark by saying that they do not make clear sense or that they are metaphysical claims in some derogatory connotation of that word. I merely want to maintain that once all of the relevant information is taken into account, the truth-values of these claims are left undetermined, and that they therefore do not constitute adequate solutions to the problems to which they are addressed.

My reason for holding this is as follows: In order that it be determinate whether or not two terms refer to the same things, it must be determinate whether or not they are referential expressions in the first place. However, as I shall subsequently argue, the status of theoretical terms as referential or non-referential is simply underdetermined. In view of this, any claim whatsoever about the the referents or extensions of such terms must itself be inderminate.

A test for whether or not an expression is referential may be conducted in a fashion prompted by certain discussions in Quine's *Word and Object*. Quine has there noted that words like 'sake' and 'behalf', though grammatically common nouns, should not be taken to refer, because their occurrences are syncategorematic ones. That is to say, these words are best understood not as significant terms in their own right, but rather as mere parts of more comprehensive expressions not further parsable into semantically significant components. Quine does not articulate the criteria by which he adjudges 'sake' and 'behalf' syncategorematic, but reflection on the case readily yields something like the following guidelines: The earmark of genuine categorematicity in an expression is the fact that the expression occurs in the whole wide range of grammatical contexts appropriate to its grammatical classification; thus, the sign of syncategorematicity is the fact that the expression occurs only in a limited number of special types of context.

Consider 'behalf': It occurs in expressions of the form 'on x's behalf', 'on the behalf of x', 'in x's behalf', 'in the behalf of x', and only in these. Because it occurs in these contexts, contexts peculiar to common nouns and noun phrases, we are able to classify it as to grammatical type. As a common noun, it is *prima facie* a referring expression. Because it occurs only in these four types of context, and not in such further contexts appropriate to common nouns as 'x is a behalf', 'behalves are F', 'the behalf that is F is G' and so on, it does not pass the test for categorematicity, is rather syncategorematic, and hence is not a genuine referring expression.

The test here proposed needs to be carefully distinguished from Quine's well known test for ontic commitment. "To be is to be the value of a variable" purports to apply to formalized or to canonical discourse and to specify what entities and kinds of entities there have to be in order that the assertions in this discourse count as true. The test I am proposing, on the other hand, applies to ordinary discourse, discourse that is amenable to canonical paraphrase and thence to formalization, but on which such paraphrasis has not yet been performed. The question to which it addresses itself may be cast in terms of Quine's criterion as follows: What are the features of ordinary discourse that require us to countenance entities it may seem to be about as values of our variables in a formalized version of that discourse? The answer has three parts. (1) An expression of the discourse is categorematic just in case it occurs in the whole wide range of grammatical contexts appropriate to expressions of its grammatical type. (2) An expression is referential just in case it is both categorematic and a grammatically referential type of expression. (3) Our variables ought to range over just those entities that are needed as referents of referential[2] expressions in order that the discourse come out true. What we have here then is no rival to Quine's criterion, but rather a needed supplement – a test for the referentiality of putatively referential terms in ordinary discourse.

Now in order to determine, along these lines, whether or not (say) 'kinetic energy' is best understood as a referring expression in Newtonian mechanics, we need first to determine its grammatical type, and then whether or not it occurs, in that theory, in the whole range of contexts appropriate to expressions of that grammatical type. If doing this were a straightforward matter, we could straightforwardly settle the question of the referential status of the term, 'kinetic energy'. However, I shall argue that it is no straightforward matter at all.

To make the decision possible, we need first of all to find out just which sentences containing the term 'kinetic energy' count as parts of the theory proper. That expressions like 'x is kinetic energy', 'kinetic energies that are F', and so on, occur in sentences uttered by persons familiar with the theory, persons talking informally in terms of the theory, may merely reflect an inclination or prejudice on the part of these persons in favor employing 'kinetic energy' as a referring expression. In checking for its referential status *in* the theory, we need thus to distinguish such sentences from sentences which are actually part of the theory.

Once such delimiting of the relevant body of discourse is accomplished, the variety of sentences that remain is severely limited. The mere fact that no set of sentences in the theory will (we hope) contradict any set of sentences in the theory rules out many sentences. Other limiting factors include the fact (given the nature of theories) that universal generalizations will predominate, while singular terms and therefore the sorts of constructions peculiar to them will tend to be rare. There will also be few, if any, existential generalizations. To note all of this is merely to recognize that a scientific theory is not a language, but contains only a small subset of the sentences of some language. The effect of this fact on our ability to determine whether 'kinetic energy' is a referential expression or not, is, of course, profound. The tests for categorematicity which require that we first identify the grammatical type of the expression and then determine whether it occurs in all the contexts appropriate to that grammatical type are tests to be performed on a language and not on some limited subset of its sentences. Given nothing but the limited variety of sentences of the theory, we may well be unable to determine categorematicity, for if we find that the expression fails to occur in certain types of context, we cannot distinguish the case where it does not occur in those types of context anywhere in the language, and is therefore syncategorematic, from the case where it does occur in such contexts but only in sentences outside the theory, and is therefore categorematic. We may even be unable to determine the grammatical type of the expression. For example, if 'kinetic energy' occurs only in general-term-type contexts in sentences of the theory, then its grammatical type may be that of an ordinary common noun phrase or that of a mass term, the former if the rest of the sentences of the language employ it only in general-term-type contexts, the latter if it also occurs in singular-term-type contexts. To apply the test, then, we need to go beyond the sentences of the theory proper, to the other sentences of the language in which the theory is couched.

Thus far, my contention that the referential or non-referential character of theoretical terms like 'kinetic energy' is underdetermined amounts simply to the observation that appeal to the sentences of the theory leaves this character underdetermined. It might appear, though, that we need merely consult the remainder of the language of which the theory is a part to settle the matter. I wish to argue, however, that there is no unique language of which a given theory is a part, and hence that the prospect of settling the matter in the way just suggested is illusory.

Suppose that a full language, L_1, be proposed as the language in which classical mechanics is couched. And suppose that a second, quite different language, L_2, is likewise proposed for this role. That it is possible to propose two such different languages is a part of the upshot of the underdetermined status of many of the expressions in the theory. That is to say, we can eke out the sentences of the theory by the addition of alternative sets of further sentences. If one prefers to think of a language, not as the set of all sentences, but rather as rules determining the sentences, the same point can be made by speaking of rules. Given a limited set of sentences, there are alternative lists of rules that will count all of those sentences as well-formed, but that will count different sets of additional sentence as also well-formed. How now should we decide which of L_1 and L_2 is the real language of classical mechanics? I ignore here for simplicity's sake the prospect that both are wrong and that some other alternative, L_3, is the right language.

First I should like to point out that the theory is the dog and the language the tail in this connection. By this I mean that the whole *raison d'être* of the language lies in the theory, that it is for the sake of supporting the theory, and only for this reason, that the language is needed at all. To observe this is merely to note that apart from our embracing or entertaining classical mechanics, or some alternative theory such as the relativistic, we should have no occasion to use expressions like 'kinetic energy' at all. Thus the sole criteria by which to assess L_1 and L_2 (apart from narrowly logical ones which I assume to be fulfilled ubiquitously here) are criteria bearing on the adequacy of the language in meeting the demands of the theory. Does the language account for, or is it at the least compatible with, the functioning of the theory? This is the sort of question to be raised.

But in raising it, we encounter a certain circularity. For we do not independently know the detailed functioning of the theory. 'Does the theory make reference to kinetic energy or does it not?' is a question about its function, yet it is precisely this kind of question we have to settle by going beyond the theory to the language as a whole. The undetermined status of certain expressions in the theory renders equally underdetermined what language we are to take to be the language of the theory. We can never decide between L_1 and L_2 precisely because the theory allows us a certain leeway in selecting a language to couch it in. The demands made by the theory are not stringent enough uniquely to determine some language as *the* language of the theory.

To summarize briefly then the import of my remarks thus far: For an expression to be a referential one is for it to play a certain sort of grammatical role in the language to which it belongs. Expanding on some suggestions of Quine, I have tried to indicate just what that role is and hence what grammatical behavior counts an expression as referential. Scientific theories do not uniquely determine the languages in which we must take them to be couched and therefore do not determine the referential status of some of their expressions. So anyone who elects to regard such expressions as referential ones has gone beyond the demands of the theory in an arbitrary way; similarly anyone who refuses these expressions referential status. It is to this situation that I wish to point in saying that sentences about the references of theoretical terms are neither true nor false.

I need now to qualify the position that I have been developng in order to account for the fact that not every expression in a scientific theory turns out to be indeterminate with respect to categorematicity. Let me return to 'behalf' to examine some further features of it. To say that 'behalf' is a syncategorematic expression is to presuppose that there are more comprehensive expressions in which it occurs that are categorematic. And, as a matter of fact, there must be. We are guaranteed that by successively attending to more and more comprehensive expressions we will eventually come to something that passes our test, simply because we must eventually come to a whole sentence. All sentences occur categorematically by dint of what we count as a grammatical context. A grammatical context of an expression is at most the rest of the sentence in which the expression occurs. Thus whole sentences qualify as categorematic as soon as they occur at all.

As we pass from 'behalf' to 'on x's behalf' to 'y Fs (does something or other) on x's behalf', we come closer and closer to a whole sentence. 'y Fs on x's behalf' is already of the form of a whole sentence and therefore a categorematic form just in case we can justify the variables and the dummy verb-constant, 'F'. And we justify these by showing that every substituend they allow can actually occur. We need not have found this, and might have to replace some of the variables with constants to reach categorematicity. Consider the full sentence, 'Heavens to Betsy'. If we also find in the language 'Heavens to Fred', 'Heavens to Alice', 'Heavens to someone', 'Heavens to everyone', *et al.*, we are justified in regarding 'Heavens to x' as categorematic. Otherwise we shall have to take the whole sentence as not susceptible to parsing into further semantically significant components. But even then, 'Heavens to Betsy' itself will count as categorematic as a whole.

What these observations reveal is that even given only a limited set of sentences, we can identify certain categorematic expressions. If the sentences in the set are all and only those sentences that are systematically related to one another in a certain way, as in the case of scientific theories, then we shall probably not have to extend all the way to sentences to find categorematicity. The more comprehensive the expression we examine, the more limited will be the variety of contexts appropriate to expressions of that grammatical type. Thus, the probability becomes greater that we may find the expression occurring in the whole range of appropriate contexts.

Consider 'kinetic energy': Although the categorematicity of this expression may be indeterminate, more comprehensive expressions such as 'x has kinetic energy' or 'the kinetic energy of x is y' may turn out to be categorematic even in terms of the limited sentences we have to go on. I have brought this matter up partly just to set the record straight, by indicating that it is fully consistent with the account I have given that many expressions in a scientific theory should be determinably referential, our inability to appeal to the entire language notwithstanding. But I have also brought it up in order to indicate that it cannot be construals of this kind that Professor Putnam intends to assign to theoretical terms.

If in referring to the extension of 'kinetic energy' one means to refer to the extension of 'x has kinetic energy', where 'x' ranges over physical objects, then the extension in question is the very same set as the extension of 'x has velocity', 'x has momentum' and a host of other expressions of the theory. For the physical objects that have kinetic energy are just those that have velocity, just those that have momentum and have a number of other physical properties. To say that 'kinetic energy', construed in this way, retains that same extension through a certain theory change is merely to say that it remains a member of a very large set of physical expressions quite distinct in meaning but applying to the same entities. Matters get even worse if we allow objects whose kinetic energy is (say) zero ergs to count as having kinetic energy, on the ground that to say that their kinetic energy is zero ergs is to say that they have a kinetic energy, namely zero ergs. For now the term has universal extension among physical objects and comes to be coextensional not only with the terms just mentioned but even with 'x has mass', 'x has length' and most of the vocabulary of physics if construed in the same way. Claims of sameness of extension among physical predicates construed in this fashion are quite trivial, compatible as they are with virtually any physical predicate turning into virtually any other without modification in extension and so cannot, I take it, be the sorts of claim Professor Putnam is interested in making.

What of construal as 'the kinetic energy of x is y' perhaps with specification of the units of measure built into the predicate? To the extension of such a predicate there belong ordered pairs whose first members are physical objects and whose second members are numbers specifying the magnitudes of the kinetic energies of those objects on the chosen scale. Refinements that might prove necessary, such as the the addition of a third place to range over times, I shall ignore, just as I did in the earlier case, since they make no essential difference to the point to be made. That point is this: However admirably suited such ordered pairs may be, on many grounds, as referents of the term 'kinetic energy', an extension composed of them would not remain the same through any change of theory that altered, however slightly, the value to be assigned a kinetic energy in a given case. Since the change from classical to relativistic mechanics does, in fact, alter these values, it follows that if kinetic energy is construed as a relative term of the kind under discussion, its extension does not remain the same through the change in theory. If Professor Putnam holds that the reference of 'kinetic energy' does remain the same, he must not have this articulation of the term in mind.

I take it, then, that the claim that such theorietical terms as 'kinetic energy' do not change their reference through such changes in theory as the change from Newtonian to relativistic mechanics does not represent a claim about terms of the kinds we have justs considered. Yet I also take it that it is only terms of these kinds whose referential status is determined by the theory.

On the other hand, if we understand 'kinetic energy' not as a syncategorematic part of some wider expression such as 'has kinetic energy' but rather as a term in its own right, true of kinetic energies themselves, that is, a term having in its extension quantities or bundles of kinetic energy, we seem then to be talking on the same wavelength as Professor Putnam. For in thus coming to regard kinetic energy as an actual stuff, we have perhaps arrived at a construal in terms of which it can reasonably be claimed that the extension of the term remains the same through the change from Newtonian to relativistic mechanics. The differences owing to change in theory can all be regarded as mere differences in the amounts of kinetic energy associated with certain specific objects under certain specific circumstances. But no change need be supposed to have been introduced respecting what underlying stuff it is whose quantitative treatment has thus changed.

It seems highly likely then that this is the sort of construal that Professor Putnam intends. Yet it is just this sort of construal I have tried to argue is unwarranted. It is precisely expressions of this kind that I have attempted to show to be indeterminate with respect to categorematicity and hence with respect to reference.

Now, if it is really true that the referential status of 'kinetic energy' is underdetermined by physical theory, then some cost should be expected to attach to arbitrarily treating the expression as referential and thus reifying kinetic energy. And we would expect this cost to turn up in the form of statements we then come to allow ourselves whose truth-values are in principle indeterminate. I think I can quickly indicate that there is just such a cost. If beyond the objects that have kinetic energy and beyond the magnitudes of the kinetic energy they have, we also recognize kinetic energy itself, that which these objects have and that of which these magnitudes measure the amount, then it becomes incumbent on us to distinguish pairs of cases in which we have the very same kinetic energy from pairs of cases in which we have kinetic energies which, though equal in amount, are not the same. If a perfectly elastic body is bouncing up and down on a perfectly elastic surface, we may then inquire, 'Is the body's kinetic energy at a certain point during bounce n the same kinetic energy as its kinetic energy at this point during bounce $n+1$ or are these different energies merely the same in magnitude?'. If it is not just that we do not know, but rather that the question does not make physical sense in terms of the theory, then we can trace this unfortunate fact to our arbitrarily overstepping the dictates of theory in permitting it to arise. Since it cannot arise if we employ only the determinately referential term, 'the kinetic energy of x is y', it is the arbitrary decision to treat 'kinetic energy' itself as referential that is responsible for the intrusion of this bit of physical nonsense.

Now, in employing the criteria for referentiality that I have, I may seem simply to have ignored the causal theory of reference proposed by Professor Putnam. Perhaps one ought to regard that theory as presenting an alternative to the grammatical theory of referentiality I have used. But I think not. The causal theory of reference is designed to answer the question what the referents *are* of referential expressions (or at least of *some* referential expressions). As such it presupposes the referentiality of physical magnitude terms rather than arguing for it.

To appreciate this I think one need only consider, one last time, the case of 'behalf'. The kinds of events that can serve as introducing events for the term 'behalf' are those on which behalves are causally efficacious. For example, the day before yesterday, I did something in my wife's behalf. That is to say, her behalf caused a certain action of mine. For a young child who might have been present, and to whom I might have explained what was going on, this could have served as an introducing event. Afterwards, we could have been confident that he was indeed making reference to my wife's behalf when he spoke of it. Why not? Only because 'behalf' is not a referring expression, a fact that the causal theory is simply not designed to cover.

Washington University

NOTES

[1] At this point, when I delivered my comments at the Symposium, I inserted with reference to Shapere and Engels the parenthetical aside, 'an amusing juxtaposition'. Professor Putnam, in his response, seized upon this remark as evidence of my 'anticommunism'. I did not then get the chance to reply, and so I should now like to set the record straight. I find Putnam's inference a pretty loose one, and should think it at least equally plausible that I was being, not anticommunist, but antiShapere.

[2] I am here and throughout using 'reference' and its cognates to embrace both the members of the extensions of general terms and the denotata of singular terms.

N. L. WILSON

ON SEMANTICALLY RELEVANT WHATSITS:
A SEMANTICS FOR PHILOSOPHY OF SCIENCE

I

We shall take it that whatever a language is, it somehow "consists" of its phonology, its (grammatical and logical) syntax and its semantics, in the sense that once the phonology, syntax and semantics are given, that constitutes a sufficient characterization or definition of the language. Here we shall be concerned, for the most part, with the question: How does one *give* the semantics of a language? I shall argue that even if there were meanings, whether reified or somehow not reified, a meaning is not a semantically relevant whatsit – which is to say "meanings" are not what you give in characterizing the semantics of a language. As a consequence, the question, "Change of meaning or change of belief?" either falls or has to be recast.

Now the question as to the semantics of a language turns into the question as to the semantics of its terms. It will be instructive to deal with just three cases, proper names ('Zeno'), logical signs ('and'), and predicates ('red').

Suppose you and I are talking about Zeno (*i.e.*, using 'Zeno'). Provided we are both talking about Zeno the Eleatic or both talking about Zeno the Stoic then, other things being equal and barring pathological breakdowns of communication, we will understand one another, we will succeed in exchanging information – as would not be the case if one of us were talking about the Eleatic and one the Stoic. If we are talking about the same man then the word 'Zeno' "has the same significance" – as I shall say – for each of us. More generally, you and I use the same language, if and only if we have the same vocabulary, the same sentences, and every word of that vocabulary (and, more generally, complex expressions too) "has the same significance" for each of us. And the point of the example is to show that the significance of the word 'Zeno' is the man, Zeno (whichever one he is). More generally, the significance of a proper name *is its nominatum*. The individual named is the semantically relevant whatsit. In characterizing English we would say: In English, 'Socrates' signifies (or, *G*-designates, as I sometimes say) the man, Socrates. (Having used 'Zeno' to make my point I shall drop it because I want to keep away from the matter of ambiguity.)

But of course we take a different line in the case of descriptions. 'The Greek philosopher who invented the paradoxes of motion' does not signify Zeno the Eleatic. It is significant by being composed in a significant way out of significant components. The man, Zeno, is not the semantically relevant whatsit of the description. (The traditional distinction between meaning and reference does not cut at the joints and the whole notion of a semantics of meaning and a semantics of reference is, I should think, on the wrong track.)

The second case. It seems plausible to suppose that you and I use the word 'and' "with the same significance" just in case we both attribute to it the same syntactical properties, just in case, *e.g.*, for both of us 'and' is a binary sentential connective and a full sentence of 'and' logically implies each of its sentential components and is logically implied by them jointly. More generally, our logical signs have-the-same-significance just in case we have the same logical truths and recognize the same logical inferences. Hence (to oversimplify a bit) the semantically relevant whatsit for a set of logical signs taken *en bloc* is the set of logical truths – specified by some enumeration or other and by recursion. Whether all this should be regarded as part of semantics or part of logical syntax does not matter.

The case of 'red' is somewhat puzzling. Obviously we cannot lay down axioms for it as we might for 'and', so there is nothing for it but to treat it like 'Zeno'. You and I use the word 'red' with the same significance just in case we both use it to signify the same property, redness, presumably. In short the semantically relevant whatsit for a *primitive* first degree predicate – adjective or common noun – is the property it signifies. In describing English we say: In English 'red' signifies the property, red (or, redness, it does not matter). I have found that people generally get all hot and bothered over the simple-minded platonism adopted here and they want to propose alternatives. In every case up to now I have found that I can show that the alternative suggested would not work. You cannot conjure with truth, the class of red things or with 'describes it as red'.

It has been suggested by some (e.g., Carnap) that the property is the meaning or intension of the predicate in question. This cannot be so. In analogy to 'The morning star = the evening star' let us consider 'red = the color of ripe strawberries'. In Frege's example we note that since the identity sentence is true, there is just one referent. But since it is not necessarily true the arguments do not have the same meaning, and if there are meanings, the meanings must be distinct from the referent. *Pari passu*, we note that the second example is true and hence there is only one referent, the property, red, and since it is not necessarily true the two meanings (if there are meanings) must be distinct from the property, red. Hence red is not a meaning. Properties generally are not meanings any more than individuals are. I do not believe it possible to prove that there are not (intensional) meanings any more than it is possible to prove that there are not gremlins. (Not that I have not tried.) But I do claim to have shown, by implication, that a meaning, supposing there to be such a thing, is not a semantically relevant whatsit.

There are systematic reasons, which I will not go into, for recognizing atomic propositions, e.g., the proposition that Socrates is pale. This is not a meaning, since its constituents are the man, Socrates and the property, *pale*, neither of which are meanings. I sometimes speak of G-propositions to underline the contrast between them and the more fashionable intensional entities (which I do not believe in anyway).

We may assume that there are some predicates of English (e.g., 'bachelor') for which there are analytic definitions. We are forced to deny that such predicates signify properties. For if there were the property, bachelorhood, then I do not see how we could avoid recognizing the (false) proposition that Socrates is a bachelor, and the molecular proposition that Socrates is never-married and Socrates is adult and Socrates is male and Socrates is human. And, recognizing molecular propositions, I do not see how we could avoid recognizing general propositions, e.g., the proposition that something is Scotch and is identical with all and only those things which are authors of *Waverley*, which is presumably identical with the proposition that the author of Waverley is Scotch, which in turn is identical with the *atomic* proposition that Sir W. Scott is Scotch. But that is absurd. So we had better dump all but primitive properties.

But what properties are there? For a start, we may take it that all observable qualities (red, hotness), all biological species (man, tiger, plantain) and all matterkinds (water, gold) are properties. I think we have to take 'H_2O' or 'the lowest oxide of hydrogen' as a description. Hence:

water = H_2O, on a par with: Scott = the author of *Waverley*. I think we have to take 'californium' as an abbreviation of 'the ninety-eighth element' which is to be regarded as a *fulfilled* description. Hence 'californium = californium' is alright. The point is, many properties will have no names, which hardly comes as a surprise.

In the case of a defined expression one can talk about its *meaning*, as one cannot in the case of 'red'. There is no objection to: 'bachelor' means 'never-married, adult, male' provided that 'means' is read as 'means-the--same-as'. In the interests of terminological uniformity, I myself would prefer to use 'has-the-same-significance as' – hyphens being used to make a unitary two-place predicate and thereby avoid committing ourselves to *something* which is the significance of the expressions. The point is that here we have a legitimate use of 'means'. But it is not a use, obviously, that will give us a generally adequate apparatus for semantics.

II

A few pages back I expressed some doubts about the "semantics of reference" (individuals, classes and truth values) and the "semantics of meaning" (individual *concepts*, intensional properties and intensional propositions). What we have here is an ontology of individuals, G-properties and G-propositions.[1] But if the semantics of meaning is going to be thrown out, something will have to be put in its place. I propose to distinguish between primary semantics, which deals with (*inter alia*) how you give the semantics of a language (see above), and secondary semantics (or perhaps theoretical descriptive semantics), which deals with the matter of arriving at the semantics (as above) of the language used by so-and-so. That is to say, secondary semantics deals with the procedures for passing from an output, or from the speech behaviour of a "native", as they say, to a description of the language which is being used in that speech behaviour.

A few examples. Suppose a person says, "Zeno was born about 490 B. C. in Agrigentum, he was originally a Pythagorean but studied with Parmenides and was converted to the Eleatic philosophy. He devised a large number of paradoxes to combat Pythagoreanism, a few of which survive...." Our problem is one of partial decipherment: given that we know the significance of everything else in the corpus of utterances except that of 'Zeno', what significance shall we attach to 'Zeno'? Here, obviously, the word 'Zeno' signifies Zeno the Eleatic. The rule for decipherment is this: choose that entity as designatum (significance) which makes more of the assertions true than would the choice of any other entity.[2] One of the assertions happens to be false (Zeno was not born in Agrigentum) but that does not matter. Zeno the Eleatic qualifies as the significance of 'Zeno' *in that corpus* just because the assertions fit Zeno the Eleatic, not exactly, but better than they fit anybody else.

We may think of the assertions of the corpus as defining a "sense" of the word 'Zeno'. Relative to two corpora which are not logically equivalent, the word 'Zeno' will have different senses, even though they might have the same significance. But of course the sense is not a semantically relevant whatsit. Moreover we may think of the speaker as having a "concept" of Zeno which may be treated as a function of the speaker's assertions containing 'Zeno', or a function of the speaker's "direct" beliefs about Zeno. (There are difficulties in the word 'about'. I use 'direct' in such a way that my belief that Zeno was a pupil of Parmenides and Tegucigalpa is the capital of Honduras is not a "direct" belief about Zeno.) It is clear that to "grasp the significance" of the name is simply to have some beliefs expressible in sentences containing that name, these sentences as a whole fitting some one individual better than they fit anything else. Thus, to put it a bit loosely, to understand 'Zeno' is just to know something about Zeno. This will hold also of predicates. There is no question of our having an *a priori* grasp of the "meanings" of either proper names or of predicates.

I myself have a fragmentary concept of Zeno the Eleatic, but Professor Emil Barnhardt, who has devoted a lifetime to the study of Zeno has a different, because much fuller, concept of Zeno (a fuller set of beliefs about Zeno). Nevertheless we may understand each other in talking about Zeno – the only requirement being that we be talking about the same thing. The point is the difference of concept or the difference in the "sense" of the word for each of us does not preclude our understanding each other provided the significance (in this case the concrete individual) is the same. Hence, as before, the sense is not a semantically relevant whatsit; the significance is. We may speak of the "trans-theoretical" (or "trans-doxastic"?) significance of the word – to borrow half a term from Shapere and Putnam. (See above, p. 200).

It is much the same with properties. There is the property gold. Or so we must suppose in order to offer a partial description of English:

'gold' signifies, in English, (the property) gold.

Shakespeare writes,

To gild refined gold, to paint the lily,
....
Is wasteful and ridiculous excess.

I imagine I understand what Shakespeare is saying because I take it for granted that his word 'gold' signifies the property gold. But our "concepts" of gold, the "sense" the word has for each of us, is almost certainly different. I believe, among other things, that gold is metalic, precious, yellowish, and element number 79. Shakespeare believes, we may suppose, that gold is metalic, precious, yellowish and is a specific against scrofula. But obviously this difference does not preclude my understanding what Shakespeare says. There is a trans-theoretical significance, namely, the property gold, and that is presumably constant. It is conceivable that if you tabulated Shakespeare's beliefs about gold, that is, tabulated the things he *would* say about gold, and deciphered the corpus according to the Principle of Charity, it might turn out that what he says fits *ormolu* better than it fits *gold*. If that were the case it would mean that all of us are simply misunderstanding the passage quoted.

Leaving aside that possibility however, we may say that gold itself is the cluster of *all* its properties – again borrowing half of Putnam's phrase "cluster concept".[3] Gold is not a conjunction of properties because most of the properties mentioned are in fact *contingent* properties of gold; and "cluster" seems to be the right word for underscoring that fact. The relation of the cluster to my concept is just that some members of that total cluster of properties are (rightly) believed by me to *be* members of the cluster. But my concept is not a concept in the sense of permitting one to read off a batch of analytic truths from it, as if my concept of gold and the (alleged) *meaning* of 'gold' were the same. (I daresay my concept of *bachelor* is something from which I *can* read off analytic truths.)

The theoretical terms of science are not so very different, but there are mplications. 'Kinetic energy'[4] is presumably definable and so does not into the present account of significance. Simply translating freely from the Greek we would have 'kinetic energy of x' means (-the-same-as) 'the energy that x has in virtue of being in motion'. This is *the* meaning – the trans-theoretical meaning, if you like. Different theories will prescribe different measures of kinetic energy but such differences issue out of differences in belief about kinetic energy, not from differences in "meaning" attached to the phrase. The view that 'kinetic energy e' at one time *meant* '$\frac{1}{2}mv^2$' is the product of a far-out operationist theory of "meaning" (which of course had a near-in point).

'Temperature', unlike 'kinetic energy', is, I should think, not definable. But unlike 'red' or 'gold' it does not signify a determinate property. It might be convenient to treat 'temperature' as a functor expression, but there does not seem to be any harm in treating temperature as a determinable property, with specific temperatures falling under it as determinates (as color is a determinable with red and green as determinates). The central facts about temperature will encapsulate the central facts about the determinate temperatures. A corpus of utterances giving a sense to 'temperature' will include theoretical, factual sentences from thermodynamics, statistical mechanics and what have you, and also the factual sentence, say, 'High temperatured objects are hot and low temperatured objects are cold', hotness and coldness being observable properties. The significance of 'temperature' is simply the determinable property, and that is trans-theoretic.

We may summarize by applying the present apparatus to the topic of this volume.

A. *Change of Meaning*

In Shakespeare's time 'fond' meant 'foolish'.[5] Nowadays it means 'affectionate or liking'. There has been a change of meaning. But assuming that neither 'foolish' nor 'affectionate' are indefinable, there are no properties signified and no question of my beliefs about such properties. Hence the example is of more interest to lexicographers than to philosophers.

B. *Change of Sense and Change of Belief*

They are the same. Any change in the set of my beliefs about gold – whether by addition or subtraction or both ("change of mind") *is* a change in my concept of gold, in the sense of the word 'gold' for me.

C. *Change of Significance*

It is at least conceivable that a person might start out with beliefs about Poincaré which fit Henri Poincaré better than they do anybody else but end up with beliefs that fit Raymond Poincaré still better. (I have met people who believe that there was just one man, the great French mathematical politician.) This would be a change of significance. Similarly it is conceivable that at one time 'mandrake' should have signified the mushroom, *amanita muscaria*, but by reason of change of belief, later came to signify the herb, *mandragora officinarum*.[6] Again, we may have beliefs about electrons such that by 'electron' we signify the property of being a particle with such and such characteristics. But there may be a marked change of belief so that we end up using 'electron' to signify a different property, the property of being an undulation of such and such characteristics. (There would also be a conviction that the first property is unexemplified.) Changes in significance must be quite rare – if indeed they occur at all. But the theoretical possibility of such shifts is perhaps of some philosophical interest.

D. *Some General Remarks*

The whole thrust of the present analysis is against the theory of *a priori* entertainable meanings. It would be held that it would be theoretically possible for Athene to spring, not just fully-armed, from the brow of Zeus, but with a full *a priori* command of the Greek language. That is false. She could not have known Greek without a fair knowledge of the world of the Greeks. (I daresay such knowledge could have been miraculously imparted to her *in capite*, but that is a different matter.)

I think, too, we have to down-grade the importance of rules in the philosophy of language. The notion of a rule has a limited relevance in the area of grammatical and stylistic correctness. If I restrain myself from splitting an infinitive or from saying "between you and I", I am perhaps following a rule. But if I use 'Plato' to refer to Plato or use 'red' to describe something as red, that's just something I *do*. I am not following a rule; I am not restraining myself from anything. It is something I just do, in the same way that swimming is something I just do, or using a knife and fork. Some people's knife-and fork behaviour is indeed rule-governed: they make a point to avoid using the left hand to carry food from plate to palate. But most people simply shovel it in. Although their behaviour is patterned or regular they are not following any rule. And most of our speech behaviour is like that. To describe it as rule-regulated behaviour is simply not helpful. Moreover, if you think of "In English, 'Plato' designates Plato" as a *rule*, then it begins to have an air of pointlessness, of silliness that it does not have if regarded as simply a statement in English about English. I would concede, incidentally, that our use of a second language may indeed be rule-governed in a way that the use of our first is not.

III

The theory of meaning sketched in Hilary Putnam's contribution to the present volume is in many respects similar to my own semantic methods. It is, I think, insufficiently radical and its being insufficiently radical is why there is a certain oscillation between meaning and reference. He ends up with what might be called quasi-intensions. At one point he writes above, p. 204), "... different speakers use the word 'electricity' without there being a discernible "intension" that they all share. If an "intension" is anything like a necessary and sufficient condition, then I think that is right. But is does not follow that there are no ideas about electricity which are in some way linguistically associated with the word." The collection of such associated ideas is what I am calling a quasi-intension and a quasi-intension is not a semantically relevant whatsit. For A, B and C can communicate about X even if the set of their *common* ideas is too thin to single out one property among many. All that is required is that the set of A's beliefs, the set of B's beliefs and the set of C's beliefs each be rich enough to pick out a property and the property be the same in each case.

On the other hand, Putnam's account is, in another respect, a little *too* radical for my taste. I mean the "causal theory of meaning", and the importance attached to "introducing events". All this I would put down as psychologism in semantics, of which more in a moment.

The causal theory of meaning is taken over from Kripke. It does, I suspect, genuinely entail a doctrine of individual identity across all possible worlds. In any case one is entitled to conjecture that the causal theory is designed to make sense out of and support the latter doctrine. And the latter doctrine is false. There is, I should think, a sequence of possible worlds, beginning with the actual world, each possible world differing slightly from its predecessor, and such that across that sequence Julius

Caesar "changes" from being a Roman general to being an obscure Latin peasant. And there is another sequence, starting from the actual world and having the same last possible world as the first sequence does, and such that across the sequence Marc Antony "changes" from being a Roman general to being *the same* obscure Latin peasant mentioned in connection with the first sequence. Thus identity of individuals across possible worlds is not transitive, which is to say it is not indentity at all. If the doctrine of pan-trans-world indentity is false we may reasonably be dubious about a theory designed to support it – if indeed the causal theory of meaning has that purpose.

According to the causal theory, "... the knowledge an individual user of a language has need not at all fix the reference of the proper names in that individual's idiolect; the reference is fixed by the fact that that individual is causally linked to other individuals who were in a position to pick out the bearer of the name." (Above. p, 207.) This strikes me as false. Suppose someone says, "Aristocles was born in 427 B.C. in Athens. He was the son of Ariston and he had two brothers, Glaucon and Adeimantus. Aristocles studied with Socrates and after the latter's death devoted himself to the pursuit of philosophy and to the writing of philosophical dialogues celebrating the memory of his teacher. Aristocles made three trips to Syracuse in the hope of making a philosopher-king out of Dionysius II, but gave up in disgust because Dionysius would not get his assignments in on time. Aristocles died in 347 B. C.."

Given this corpus of utterances and given what we know about Greek philosophy we have no alternative but to conclude that, for this speaker, the referent (significance) of 'Aristocles' is the man, Plato. The introducing event, the causal chain allegedly linking the speaker with Plato-knowing 'Aristocles'-users is irrelevant. It does not alter the situation at all. In one sense, referring (when it holds) can be looked on as a relation between the name, a corpus of utterances in which the name occurs, and the actual world – which supplies a referent which more or less "fits" the corpus. The foregoing is offered as a typical (and therefore philosophically interesting) example of the way reference is fixed. The Kripke-Putnam account has a certain suggestiveness when it comes to degenerate cases. If we are deciphering a very thin corpus containing 'Sir William Hamilton' it will indeed make a difference whether the corpus contains, 'I first learned about Sir W. Hamilton from reading a history of Scotch philosophy' or

'... from reading a textbook on quaternions' or '... from reading a biography of Horatio Lord Nelson'. But even here, what fixes reference is *what the man says*, not how, psychologically, he comes to say it.

I must confess incidentally to a certain personal discomfort over Putnam's 'elm' and 'aluminum' examples. But even Putnam is prepared to say, "Botanists apply the term 'elm' to elms", " (American) metallurgists call aluminum 'aluminum'." And these corpora, thin as they are, are sufficient to fix the reference.

Keep psychology out of logic, we all say. And I think we should add: and keep it out of semantics, too. Semantics should be – one wants to say – *purely formal*, but that would not do. So we shall say: semantics should be *purely semantical*. Logic is formal in the sense that given a set of axioms and one or more deduction rules the theoremhood or non-theoremhood of any sentence of the system is thereby determined. The psychology of the logician or axiom producer is irrelevant. And in the same way, given a sufficiently comprehensive corpus of utterances *and* the state of the world, then (supposing the logic of the language is somehow fixed) the significance of all the primitive descriptive words is determined, together with the truth and falsity of all the sentences. An omniscient being could tag each sentence as true or false without any reference to what goes on in the producer of the corpus.

If one does not resolutely keep psychology out of semantics there is a danger of winding up with mysterious intentional or mental acts by which the speaker somehow manages to reach out and lasso the object of reference. There may or may not be mental acts, but referring or signifying is not one of them. Consider the little lecture on Aristocles. Suppose the speaker finishes by saying, "Now you all thought I was talking about Plato. I was not. I was telling you a cock and bull story about an obscure Attic peasant named 'Aristocles', whose historical reputation I was concerned to boost." I contend that the man cannot truthfully say this. There is no mental act he can perform, no mental act he can fail to perform, which will alter the referent as determined by the corpus and the facts in conjunction.[7] I will put the matter bluntly: people do not signify (or mean) things; their words in their corpora do. There is a sort of categorical imperative here: If you try to tell too big a whopper about X, you may end up telling some truth about Y, or saying nothing at all. Even if there were not arguments against mentalistic semantics, we might reasonably shrink from it. For any philosophical explication issuing from it would be *obscurum per obscurius*.[8] Putnam, to be sure, does not propose getting bogged down in that particular morass. But the causal theory of meaning is another garden path, just as it was when Ogden and Richards proposed it nearly half a century ago. Quine's theory of stimulus meaning, incidentally, is another.

But sooner or later psychology is going to rear its ugly head. For when we reverse the process of abstraction and refer the corpus of utterances back to the speaker then, having deciphered, we will have psychological statements like "So-and-so uses 'Aristocles' to signify Plato". And the psychology for philosophy of language will be, in an attenuated sense, behaviouristic. For in the concrete case, the corpus is a behavioural output, or at least the output a subject *would* produce if pressed by the demand (say), "Tell us all you know about Aristocles." It is a behaviourism that has nothing to say about language acquisition, however.[9]

Now Putnam might find behaviourism just as odious as the idealism he attacks in his paper,[10] and he may have political reasons for so finding it. When he writes (above, p. 214), "Social practice is the test of truth," I am probably not the only reader to experience a certain chill. For it is only a short step to: Truth is what the Communist Party (or the New Left, or the Pentagon) finds it expedient to regard as the truth. If the test is taken generally as a criterion for philosophical truth or correctness one is tempted to retort that political considerations have no place in such essentially frivolous and intellectual fields as philosophy of language, philosophy of science and philosophy of mind. Yet ideas *do* have consequences. And so we are inclined to imagine ourselves as philosopher-kings distilling the recondite philosophical truths which will provide the illumination and guidance so necessary for political theory and practice. But, human nature being what it is, things never work out that way. It may be that abstract theories are never really scrutinized until their bad consequences force the scrutiny. A philosopher may be an adamantine determinist until he observes that people use determinism to escape responsibility, to dehumanize themselves and others. Only after noting the consequences (if even then) will he feel impelled to go back and find the *purely intellectual* flaws in his position. *That*, it seems to me, is the relationship between the Cave and Cloud Number Nine.

McMaster University

NOTES

[1] I accept the *symbolism* of sets but hold that sets are pseudo-entities. See Wilson (1959a), Chapter V for a criterion whereby to decide whether ineliminable quantification does or does not commit us to an ontology of the apparent quantified-overs.
[2] In Wilson (1959b), I referred to this as *The Principle of Charity*.
[3] Putnam (1963), esp. p. 378.
[4] Putnam's example from the *op. cit.* I am indebted, incidentally, to Michael Radner fo warning me that 'in virtue of its motion' would not do in my definition. (Momentum Velocity?)
[5] I am indebted to Norman Rosenblood of the McMaster University English Department for the example.
[6] The example is inspired by a somewhat skeptical reading of John Allegro's *The Sacred Mushroom and the Cross*. He does suggest that in some languages some shifts of that sort have occurred. Metaphorical shifts – as from *mushroom* to *phallos* – would not, of course, serve as an example of what I have in mind here.
[7] Or consider Putnam's example, "Quine is a Roman emperor". We feel like replying, "You've got be to kidding. You must be thinking of Quintilian. But even *he* wasn't a Roman emperor."
[8] See Wilson (1970).
[9] In 'The Two Main Problems of Philosophy and Some Tips on Their Solution', given at the meeting of the Canadian Philosophical Association in St. John's, Nfld., June 9, 1971, I came out flatly against (doctrinal) logical behaviourism. Yet I think that in practice we should push behaviourism as far as it will go, just to see exactly how far it *will* go.
[10] Chomsky certainly finds it odious.

BIBLIOGRAPHY

Putnam, Hilary, 1962, 'The Analytic and Synthetic', in *Scientific Explanation, Space and Time* (ed. by Herbert Feigl and Grover Maxwell), *Minnesota Studies in the Philosophy of Science*, Vol. III, Minneapolis.

Wilson, N. L., 1959a, *The Concept of Language*, Toronto.

Wilson, N. L., 1959b, 'Substances Without Substrata', *Review of Metaphysics*, **12**, 521–39.

Wilson, N. L., 1970, 'Grice on Meaning: The Ultimate Counter-Example', *Nous* **4**, 295.

GENERAL BIBLIOGRAPHY

Achinstein, P., 'Models, Analogies, and Theories', *Philosophy of Science* **31** (1964) 328–50.
Achinstein, P., *Concepts of Science, A Philosophical Analysis*, Baltimore 1968.
Achinstein, P., 'On the Meaning of Scientific Terms', *Journal of Philosophy* **61** (1964) 497–509.
Achinstein, P., 'Theoretical Models', *British Journal for the Philosophy of Science* **16** (1965–1966) 102-20.
Achinstein, P., 'Theoretical Terms and Partial Interpretation', *British Journal for the Philosophy of Science* **14** (1963–1964) 89–105.
Achinstein, P., 'Variety and Analogy in Confirmation Theory', *Philosophy of Science* **30** (1963) 207–21.
Ackermann, R., 'Confirmatory Models of Theories', *British Journal for the Philosophy of Science* **16** (1965–1966) 312–26.
Ackermann, R., 'Howard Kahane's Entrenchment Theory', *Philosophy of Science* **33** (1966) 70–75.
Ackermann, R., Review of 'Models and Analogies in Science' (by M. B. Hesse), *British Journal for the Philosophy of Science* **16** (1965–1966) 161–63.
Addison, J. W., Henkin, L., and Tarski, A., *Theory of Models*, International Symposium at Berkeley, Proceedings, 1963, Amsterdam 1965.
Agassi, J., 'Analogies as Generalizations', *Philosophy of Science* **31** (1964) 351–56.
Agassi, J., 'Between Micro and Macro', *British Journal for the Philosophy of Science* **14** (1963–1964) 26–31.
Agassi, J., 'Science in Flux: Footnotes to Popper', in *Boston Studies in the Philosophy of Science*, vol. III (ed. by R.S.Cohen and M.W.Wartofsky), Dordrecht, 1967 pp. 293–330.
Agassi, J., *The Continuing Revolution: A History of Physics from the Greeks to Einstein*, History of Science Series (ed. by D. A. Greenberg), New York 1968.
Aiken, H. D., 'Notes on the Categories of Naturalism', *Journal of Philosophy* **43** (1946) 517–26.
Ajdukiewicz, K., 'Logic and Experience', *Synthese* **8** (1949–1951) 289–99.
Albritton, R., 'On Wittgenstein's Use of the Term "Criterion"', *Journal of Philosophy* **56** (1959) 845–57.
Aldrich, V. C., 'Logically Necessary a Posteriori Propositions', *Analysis* **29** (1968–1969) 140–42.
Aldrich, V. C., 'Mr. Quine on Meaning, Naming, and Purporting to Name', *Philosophical Studies* **6** (1955) 17–26.
Aldrich, V. C., 'The Origin of A Priori', *Journal of Philosophy* **51** (1954) 229–37.
Alexander, H. G., 'Convention, Falsification and Induction', *Aristotelian Society, Proceedings*, Supplementary **34** (1960) 131–44.
Alexander, H. G., 'Language and Metaphysical Truth', *Journal of Philosophy* **34** (1937) 645–52.
Alexander, H. G., 'Linguistic Morphology in Relation to Thinking', *Journal of Philosophy* **33** (1936) 261–69.

Alexander, H. G., 'More about the Paradigm-Case Argument', *Analysis* **18** (1957–1958) 117–21.
Alexander, H. G., 'Necessary Truth', *Mind* **66** (1957) 506–21.
Alexander, P., 'Other People's Experiences', *Aristotelian Society, Proceedings* **51** (1950–1951) 25–46.
Alexander, P., 'Theory-Construction and Theory-Testing', *British Journal for the Philosophy of Science* **9** (1958–1959) 29–38.
Alston, W. P., 'Philosophical Analysis and Structural Linguistics', *Journal of Philosophy* **59** (1962) 709–20.
Alston, W. P., 'Ziff's Semantic Analysis', *Journal of Philosophy* **59** (1962) 5–20.
Altschul, E., and Biser, E., 'The Validity of Unique Mathematical Models in Science', *Philosophy of Science* **15** (1948) 11–24.
Ambrose, A., 'On Entailment and Logical Necessity', *Aristotelian Society, Proceedings*, n.s. vol. **56** (1955–1956) 241–58.
Ambrose, A., 'The Problem of Linguistic Inadequacy', in *Philosophical Analysis* (ed. by M. Black), Ithaca, 1950, pp. 14–35.
Anderson, A. R., 'Church on Ontological Commitment', *Journal of Philosophy* **56** (1959) 448–52.
Anderson, A. R., *Minds and Machines*, Englewood Cliffs, N.J. 1964.
Anscombe, G. E.M., 'The Principle of Individuation', *Aristotelian Society, Proceedings*, supplementary vol. 27 (1953) 83–96.
Aune, B. A., 'Is There an Analytic A Priori?', *J. of Philosophy* **60** (1963) 281–91.
Aune, B. A., 'The Problem of Other Minds', *Philosophical Review* **70** (1961) 320–39.
Austin, J. L., 'Are There A Priori Concepts?', *Aristotelian Society, Proceedings*, supplementary **18** (1939) 83–105.
Austin, J. L., 'A Plea for Excuses', *Aristotelian Society, Proceedings*, n.s. **57** (1956–1957) 1–30.
Austin, J. L., *Sense and Sensibilia*, Oxford 1962.
Austin, J. L. and Naess, A., *John Austin and Arne Naess on Herman Tennessen's Experimental Warning: 'What Should We Say?'*, Oslo 1959; Berkeley 1960.
Ayer, A. J. (ed.), *Logical Positivism*, New York 1959.
Ayer, A. J., *The Revolution in Philosophy*, London 1956.
Ayer, A. J., 'Truth by Convention', *Analysis* **4** (1936–1937) 17–22.
Ayer, A. J., 'Verification and Experience', in *Logical Positivism* (ed. by A. J. Ayer), New York, 1959, pp. 228–43.
Baier, K., 'Decisions and Descriptions', *Mind* **60** (1951) 181–204.
Baier, K. and Toulmin, S. E., 'On Describing', *Mind* **61** (1952) 13–38.
Baier, K., 'The Ordinary Use of Words', *Aristotelian Society, Proceedings*, n.s. **52** (1951–1952) 47–70.
Baldin, S. F., 'Energy as the Basic Concept for a Unified Interpretation of Physical Phenomena', *Philosophy of Science* **9** (1942) 294–305.
Bar-Hillel, Y., 'A Prerequisite for Rational Philosophical Discussion', in *Logic and Language, Studies Dedicated to Professor Rudolf Carnap* (ed. by B. H. Kazemier and D. Vuysje), Dordrecht, 1962 pp. 1–5.
Bar-Hillel, Y., *Language and Information*, Reading, Mass. 1964.
Bar-Hillel, Y. (ed.), *Logic, Methodology and Philosophy of Science*, Vol. 11, Amsterdam 1964.
Bar-Hillel, Y., 'Logical Syntax and Semantics', *Language* **30** (1954) 230–37.
Bar-Hillel, Y., 'On Syntactical Categories', *Journal of Symbolic Logic* **15** (1958) 1–16.

Barker, S. F., 'On Simplicity in Empirical Hypotheses', *Philosophy of Science* **28** (1961) 162–71.
Barker, S. F., 'The Role of Simplicity in Explanation', in *Current Issues in the Philosophy of Science* (ed. by H. Feigl and G. Maxwell), New York, 1961, pp. 265–86.
Barker, S. F. and Achinstein, P., 'On the New Riddle of Induction', *The Philosophical Review* **69** (1960) 511–22.
Barnes, W. H. F., 'Meaning and Verifiability', *Philosophy* **14** (1939) 410–21.
Barrett, R. B., 'Quine, Synonymy and Logical Truth', *Philosophy of Science* **32** (1965) 361–67.
Barrett, W., 'Logical Empiricism and the History of Philosophy', *Journal of Philosophy* **36** (1939) 124–32.
Barrett, W., 'Reduction – Sentences and Extensionality', *Analysis* **4** (1936–1937) 71–7.
Bartley III, W. W., 'Achilles, the Tortoise, and Explanation in Science and History', *British Journal for the Philosophy of Science* **13** (1962–1963) 15–33.
Basilius, H., 'Neo-Humboldtian Ethnolinguistics', *Word* **8** (1952) 95–105.
Basson, A. H., 'The Existence of Material Objects', *Mind* **55** (1946) 308–18.
Basson, A. H., 'Logic and Fact', *Analysis* **8** (1947–1948) 81–7.
Basson, A. H., 'The Logical Status of Supposition', *Aristotelian Society, Proceedings*, supplementary vol. **25** (1951) 99–110.
Basson, A. H., 'Unsolvable Problems', *Aristotelian Society Proceedings* **57** (1956–1957) 269–80.
Basson, A. H. and O'Connor, D. J., 'Language and Philosophy', *Philosophy* **22** (1947) 49–65.
Baumer, W. H., 'Evidence and Ideal Evidence', *Philosophy and Phenomenological Research* **24** (1963–1964) 567–72.
Baylis, C. A., 'Logical Subjects and Physical Objects', *Philosophy and Phenomenological Research* **17** (1956–1957) 483–87.
Baylis, C. A., 'The Nature of Evidential Weight', *Journal of Philosophy* **32** (1935) 281–86.
Beardsley, M., 'Categories', *Review of Metaphysics* **8** (1954–1955) 3–29.
Beck, L. W., 'Constructions and Inferred Entities', in *Readings in the Philosophy of Science* (ed. by H. Feigl and M. Brodbeck), New York, 1953, pp. 368–81.
Beck, L. W., 'On the Meta-Semantics of the Problem of the Sythetic A Priori', *Mind* **66** (1957) 228–32.
Beck, L. W., 'The Distinctive Traits of an Empirical Method', *Journal of Philosophy* **44** (1947) 337–44.
Beck, L. W., 'Once More unto the Breach: Kant's Answer to Hume', *Ratio* **9** (1967) 33–7.
Beck, L. W., 'Remarks on the Distinction between Analytic and Synthetic', *Philosophy and Phenomenological Research* **9** (1948–1949) 720–27.
Beck, L. W., 'The Second Analogy and the Principle of Intermediacy', *Kant-Studien* **57** (1966) 199–205.
Becker, E. F., 'Indeterminacy Defended', *Philosophical Studies* **22** (1971) 1–9.
Benacerraf, P., 'What Numbers Could Not Be', *Philosophical Review* **74** (1965) 47–73.
Benjamin, A. C., 'The Concept of the Variable-Given', *Journal of Philosophy* **33** (1936) 225–30.
Benjamin, A. C., 'A Definition of "Empiricism"', *Philosophy and Phenomenological Research* **15** (1954–1955) 171–79.
Benjamin, A. C., 'Modes of Scientific Explanation', *Philosophy of Science* **8** (1941) 486–92.
Benjamin, A. C., 'On Defining Science', *Scientific Monthly* **68** (1949) 192–98.

Benjamin, A. C., 'Outlines of an Empirical Theory of Meaning', *Philosophy of Science* **3** (1936) 250–66.
Benjamin, A. C., 'Some Theories of the Development of Science', *Philosophy of Science* **20** (1953) 167–76.
Bennett, J., 'Analytic-Synthetic', *Aristotelian Society, Proceedings* **59** (1958–1959) 163–88.
Bennett, J., *Kant's Analytic*, Cambridge, England 1966.
Bennett, J., 'Meaning and Implication', *Mind* **63** (1954) 451–63.
Bentley, A. F., 'Logicians' Underlying Postulations', *Philosophy of Science* **13** (1946) 3–19.
Bentley, A. F., 'On a Certain Vagueness in Logic', *Journal of Philosophy* **42** (1945) 6–27, 39–51.
Berenda, C. W., 'On Verifiability, Simplicity, and Equivalence', *Philosophy of Science* **19** (1951) 70–6.
Berg, J., 'On the Argument Against Reduction Sentences', *Philosophy of Science* **38** (1971) 118–20.
Bergmann, G., 'An Empiricist's System of the Sciences', *Scientific Monthly* **59** (1944) 140–48.
Bergmann, G., 'Holism, Historicism, and Emergence', *Philosophy of Science* **11** (1944) 209–21.
Bergmann, G., 'The Logic of Psychological Concepts', *Philosophy of Science* **18** (1951) 93–110.
Bergmann, G., *Logic and Reality*, Madison, Wisc. 1964.
Bergmann, G., *Meaning and Existence*, Madison, Wisc. 1960.
Bergmann, G., *The Metaphysics of Logical Positivism*, London, 1954.
Bergmann, G., 'On Physicalistic Models of Non-Physical Terms', *Philosophy of Science* **7** (1940) 151–58.
Bergmann, G., *Realism – A Critique of Brentano and Meinong*, Madison, Wisc. 1967.
Bergmann, G., 'Remarks Concerning the Epistemology of Scientific Empiricism', *Philosophy of Science* **9** (1942) 283–93.
Bergmann, G., 'Sense and Nonsense in Operationism', *Scientific Monthly* **79** (1954) 210–14.
Bergmann, G., 'On Some Methodological Problems of Psychology' in *Readings in the Philosophy of Science* (ed. by H. Feigl and M. Brodbeck), New York, 1953, pp. 627–36.
Bergmann, G., 'Two Criteria for an Ideal Language', *Philosophy of Science* **16** (1949) 71–4.
Berlin, I., 'Logical Translation', *Aristotelian Society, Proceedings* **50** (1949–1950) 157–88.
Berlin, I., 'Verifiability in Principle', *Aristotelian Society Proceedings*, n.s. vol. **39** (1938–1939) 225–48.
Bertalanffy, L. von, 'An Essay on the Relativity of Categories', *Philosophy of Science* **22** (1955) 243–63.
Bertalanffy, L. von, 'An Outline of a General System Theory', *British Journal for the Philosophy of Science* **1** (1950–1951) 134–72.
Beth, E. W., 'Critical Epochs in the Development of the Theory of Science', *British Journal for the Philosophy of Science* **1** (1950–1951) 27–42.
Beth, E. W., 'Extension and Intension' in *Logic and Language, Studies Dedicated to Professor Rudolf Carnap* (ed. by B. H. Kazemier and D. Vuysje), Dordrecht, 1962, 64–8.
Beth, E. W., *The Foundations of Mathematics*, Amsterdam 1959.
Beth, E. W., 'Fundamental Features of Contemporary Theory of Science', *British Journal for the Philosophy of Science* **1** (1950–1951) 291–303.

Beth, E. W., 'The Relationship between Formalized Language and Natural Language', *Synthese* **15** (1963) 1–16.
Beth, E. W. and Piaget, J., *Mathematical Epistemology and Psychology*, New York 1966.
Bierstedt, R., 'The Meanings of Culture', *Philosophy of Science* **5** (1938) 204–16.
Binkley, R., 'A Theory of Practical Reason', *Philosophical Review* **74** (1965) 423–48.
Binkley, R., 'Quantifying, Quotation and a Paradox', *Noûs* **4** (1970) 271–7.
Binkley, R., 'The Surprise Examination in Modal Logic', *Journal of Philosophy* **65** (1968) 127–36.
Bittle, W. E., 'Language and Culture Areas', *Philosophy of Science* **20** (1953) 247–56.
Black, M., 'The Analysis of a Simple Necessary Statement', *Journal of Philosophy* **40** (1943) 39–46.
Black, M., 'Certainty and Empirical Statements', *Mind* **51** (1942) 361–7.
Black, M., 'Conventionalism in Geometry and the Interpretation of Necessary Statements', *Philosophy of Science* **9** (1942) 335–49.
Black, M., 'Is Analysis a Useful Method in Philosophy?', *Aristotelian Society, Proceedings*, supplementary vol. **13** (1934) 53–64.
Black, M., 'The Limitations of a Behavioristic Semiotic', *Philosophical Review* **56** (1947) 258–72.
Black, M., 'Linguistic Relativity: The Views of Benjamin Lee Whorf', *Philosophical Review* **68** (1959) 228–38.
Black, M., *Models and Metaphors*, Ithaca, New York 1962.
Black, M., 'Necessary Statements and Rules', *Philosophical Review* **67** (1958) 313–41.
Black, M., 'The "Paradox of Analysis"', *Mind* **53** (1944) 263–67.
Black, M., 'Truth by Convention', *Analysis* **4** (1936–1937) 28–32.
Black, M., 'Vagueness', *Philosophy of Science* **4** (1937) 427–55.
Blair, G. W. S., 'Some Aspects of the Search for Invariants', *British Journal for the Philosophy of Science* **1** (1950–1951) 230–56.
Blanché, R., *Structures intellectuelles, essai sur l'organization des concepts*, Paris 1966.
Bloomfield, L., *Language*, New York 1933.
Bloomfield, L., *Linguistic Aspects of Science*, Chicago 1939.
Blumberg, A. E. and Boas, G., 'Some Remarks in Defense of the Operational Theory of Meaning', *Journal of Philosophy* **28** (1931) 544–50.
Bochenski, I. M., Church, A., and Goodman, N. (eds.), *The Problem of Universals*, Notre Dame, Indiana 1956.
Bohm, D., 'Classical and Non-Classical Concepts in Quantum Theory', *British Journal for the Philosophy of Science* **12** (1961–1962) 265–80.
Bohm, D., 'On the Relationship Between Methodology in Scientific Research and the Content of Scientific Knowledge', *British Journal for the Philosophy of Science* **12** (1961–1962) 103–16.
Bohnnert, H. G., 'In Defense of Ramsey's Elimination Method', *Journal of Philosophy* **65** (1968) 275–81.
Bohr, N., 'Causality and Complementarity', *Philosophy of Science* **4** (1937) 289–98.
Bohr, N., 'Discussion with Einstein on Epistemological Problems in Atomic Physics', in *Albert Einstein: Philosopher-Scientist* (ed. by P. A. Schilpp), 2 vols., New York, 1959, pp. 199–241.
Bonifacio, A. F., 'On Analytic-Synthetic Truth – A Methodological Comment', *Journal of Philosophy* **56** (1959) 64–7.
Boone, W. W., 'The Word Problem', *Annals of Mathematics* **70** (1959) 207–65.

Born, M., 'The Interpretation of Quantum Mechanics', *British Journal for the Philosophy of Science* **4** (1953–1954) 95–106.
Born, M., *Natural Philosophy of Cause and Chance*, Oxford 1951.
Bradley, H., *On the Relation between Spoken and Written Language*, Folcroft, Pa. 1919.
Bradley, R. D., 'Geometry and Necessary Truth', *Philosophical Review* **73** (1964) 59–75.
Braithwaite, R. B., 'The Idea of Necessary Connexion', *Mind* **37** (1928) 62–72.
Braithwaite, R. B., 'Imaginary Objects', *Aristotelian Society, Proceedings*, supplementary vol. **12** (1933) 44–54.
Braithwaite, R. B., 'Models in the Empirical Sciences', in *Logic, Methodology and Philosophy of Science* (ed. by E. Nagel *et al.*, Stanford, 1962, pp. 224–231.
Braithwaite, R. B., 'The New Physics and Metaphysical Materialism', *Aristotelian Society, Proceedings*, vol. **43** (1942–1943) 203–09.
Braithwaite, R. B., 'Reducibility', *Aristotelian Society, Proceedings*, supplementary vol. **26** (1952) 121–38.
Braithwaite, R. B., 'The Role of Values in Scientific Inference', in *Induction: Some Current Issues* (ed. by Henry E. Kyburg and Ernest Nagel), Middleton, Conn., 1963, pp. 180–204.
Braithwaite, R. B., *Scientific Explanation*, Cambridge, England 1953.
Brandt, R., 'The Languages of Realism and Nominalism', *Philosophy and Phenomenological Research* **17** (1956–1957) 516–35.
Brandt, R., 'On the Possibility of Reference to Inferred Entities', *Journal of Philosophy* **35** (1938) 393–405.
Brandt, R. B. and Nagel. E. (eds.) *Meaning and Knowledge: Systematic Readings in Epistemology*, New York 1965.
Bridgman, P. W., *Collected Experimental Papers*, Cambridge, Mass. 1964.
Bridgman, P. W., *The Logic of Modern Physics*, New York 1927.
Bridgman, P. W., *The Nature of Physical Theory*, Princeton, N.J. 1936.
Bridgman, P. W., 'The Nature of Some of Our Physical Concepts', *British Journal for the Philosophy of Science* **1** (1950–1951) 257–72; **2** (1951–1952) 25–44, 142–60.
Bridgman, P. W., 'Operational Analysis', *Philosophy of Science* **5** (1938) 114–31.
Bridgman, P. W., 'The Operational Aspect of Meaning', *Synthese* **8** (1949–1951) 251–9.
Bridgman, P. W., 'Science: Public or Private?', *Philosophy of Science* **7** (1940) 36–48.
Bridgman, P. W., *A Sophicate's Primer of Relativity*, Middleton, Conn. 1962.
Bridgman, P. W., *The Way Things Are*, Cambridge, Mass. 1959.
Britton, K., 'Are Necessary Truths True by Convention?', *Aristotelian Society, Proceedings*, supplementary vol. **21** (1947) 78–103.
Britton, K., *Communication: A Philosophical Study of Language*, London 1939.
Britton, K., 'Empirical Foundation for Logic', *Analysis* **2** (1933–34) 37–42.
Britton, K., 'The Description of Logical Properties', *Analysis* **7** (1939–1940) 40–5.
Britton, K., 'Empirical Foundation for Logic', *Analysis* **2** (1934–1935) 37–42.
Britton, K., 'Epistemological Remarks on the Propositional Calculus', *Analysis* **3** (1935–1936) 57–65.
Britton, K., 'Reason and Rules of Language', *Aristotelian Society, Proceedings*, n.s. vol. **39** (1938–1939) 147–66.
Broad, C. D., 'Are There Synthetic A Priori Truths?', *Aristotelian Society, Proceedings*, supplementary vol. **15** (1936) 102–17.
Broad, C. D., *Scientific Thought*, London 1923.
Brodbeck, M., 'Coherence Theory Reconsidered: Professor Werkmeister on Semantics

and the Nature of Empirical Laws', *Philosophy of Science* **16** (1949) 75–85.
Brodbeck, M., 'Meaning and Action', *Philosophy of Science* **30** (1963) 309–24.
Brodbeck, M., 'Methodological Individualisms: Definition and Reduction', *Philosophy of Science* **25** (1958) 1–22.
Brodbeck, M., 'Models, Meaning and Theories', in *Symposium on Sociological Theory* (ed. by L. Gross), Evanston, Ill., 1959, pp. 373–403.
Brodbeck, M. and Feigl, H. (eds.), *Readings in the Philosophy of Science*, New York 1953.
Broglie, L. de, 'L'espace et le temps dans la physique quantique', *International Congress of Philosophy, 10th, 1949, Proceedings*, vol. I, Amsterdam, 1949, pp. 806–15.
Broglie, L. de, *New Perspectives in Physics*, New York 1962.
Broglie, L. de, *The Revolution in Physics*, New York 1958.
Bromberger, S., 'A Theory about the Theory of Theory and about the Theory of Theories', *Delaware University Seminar on the Philosophy of Science*, vol. 2, New York, 1963, 79–105.
Bronstein, D. J., 'The Meaning of Implication', *Mind* **45** (1936) 157–80.
Bronstein, D. J., 'What is Logical Syntax?', *Analysis* **3** (1935–1936) 49–56.
Brotman, H., 'Could Space be Four Dimensional?', *Mind* **62** (1952) 317–27.
Brouwer, L. E. J., 'Intuitionistische Einführung des Dimensionsbegriffes', *K. Nederlandse Akademie van Wetenschappen, Amsterdam. Afdeling Natuurkunde Proceedings of the Section of Sciences* **29** (1926) 855–64.
Brower, R. A. (ed.), *On Translation*, Cambridge, Mass. 1959.
Brown, R. W., 'Language and Categories', Appendix to *A Study of Thinking* (ed. by J. S. Bruner, J. J. Goodnow, and G. A. Austin), New York, 1956, pp. 247–312.
Brown, R. W., *Words and Things*, Glencoe 1958.
Bruner, J. S., Goodnow, J. J., and Austin, G. A., *A Study of Thinking*, New York 1956.
Brunner, K., '"Assumptions" and the Cognitive Quality of Theories', *Synthese* **20** (1969) 501–25.
Brutian, G. A., 'The Philosophical Bearings of the Theory of Linguistic Relativity', *Soviet Studies in Philosophy* **2** (1963–1964) 31–8.
Buchdahl, G., 'Causality, Causal Laws and Scientific Theory in the Philosophy of Kant', *British Journal for the Philosophy of Science* **16** (1965–1966) 187–208.
Buchdahl, G., *Metaphysics and the Philosophy of Science*, Oxford 1969.
Buchdahl, G., 'Semantic Sources of the Concept of Law', *Synthese* **17** (1967) 54–74.
Bull, R. A., 'MIPC as the Formalization of an Intuitionist Concept of Modality', *Journal of Symbolic Logic* **31** (1966) 609–16.
Bunge, M., 'Analogy in Quantum: Theory from Insight to Nonsense', *British Journal for the Philosophy of Science* **18** (1967–1968) 265–87.
Bunge, M., *Causality*, Cambridge, Mass. 1959.
Bunge, M., 'Causality, Change, and Law', *American Scientist* **49** (1961) 432–48.
Bunge, M., 'Kinds and Criteria of Scientific Laws', *Philosophy of Science* **78** (1961) 260–81.
Bunge, M., 'Strife about Complementarity', *British Journal for the Philosophy of Science* **6** (1955–1956) 1–12, 141–54.
Bunge, M., 'The Weight of Simplicity in the Construction and Assaying of Scientific Theories', *Philosophy of Science* **28** (1961) 120–49.
Burgers, J. M., *Experience and Conceptual Activity*, Cambridge, Mass. 1965.
Burtt, E. A., 'The Status of "World Hypotheses"' *Philosophical Review* **52** (1943) 590–601.
Butler, R. J., 'Language Strata and Alternative Logics', *Australasian Journal of Psychology and Philosophy* **33** (1955) 77–87.

Butterfield, H., *The Origins of Modern Science*, Glencoe 1957.
Butts, R. E., 'Feyerabend and the Pragmatic Theory of Observation', *Philosophy of Science* **33** (1966) 383–94.
Butts, R. E., 'On Walsh's Reading of Whewell's View of Necessity', *Philosophy of Science* **32** (1965) 175–81.
Butts, R. E. and Davis, J. W. (eds.), *The Methodological Heritage of Newton*, Toronto 1969.
Butts, R. E. (ed.), *William Whewell's Theory of Scientific Method*, Pittsburgh 1968.
Byerly, H., 'Model-Structures and Model-Objects', *British Journal for the Philosophy of Science* **20** (1969) 135–92.
Byerly, H., 'Professor Nagel on the Cognitive Status of Scientific Theories', *Philosophy of Science* **35** (1968) 412–23.
Caldin, E. F., 'Theories and the Development of Chemistry', *British Journal for the Philosophy of Science* **10** (1959–1960) 209–22.
Campbell, D. T., Segall, M. H., and Herskovits, M. J., *The Influence of Culture on Visual Perception*, Indianapolis 1966.
Campbell. N. R., *The Foundations of Science*, New York 1957.
Campbell, N. R., 'The Structure of Theories', in *Readings in the Philosophy of Science* (ed. by H. Feigl, and M. Brodbeck), New York, 1953, pp. 288–308.
Campbell, N. R., *What is Science?*, New York 1952.
Cardwell, D. S. L., 'Early Development of the Concepts of Power, Work and Energy', *British Journal for the History of Science* **3** (1966–1967) 209–24.
Carmichael, P. A., 'Limits of Method', *Journal of Philosophy* **45** (1948) 141–52.
Carnap, R., 'Empiricism, Semantics and Ontology', *Revue Internationale de Philosophie* **4** (1950) 20–40.
Carnap, R., *Introduction to Semantics and Formalization of Logic*, Cambridge, Mass. 1959.
Carnap, R., *The Logical Foundations of Probability*, Chicago 1962.
Carnap, R., 'Meaning and Synonymy in Natural Languages', *Philosophical Studies* **6** (1955) 33–46.
Carnap, R., 'Meaning Postulates', *Philosophical Studies* **3** (1952) 65–73.
Carnap, R., 'Methodological Character of Theoretical Concepts', in *Minnesota Studies in the Philosophy of Science*, Vol. I (ed. by H. Fiegl and M. Scriven), Minneapolis, 1956, pp. 38–76.
Carnap, R., 'Modalities and Quantification', *Journal of Symbolic Logic* **11** (1946) 33–64
Carnap, R., 'On Some Concepts of Pragmatics', *Philosophical Studies* **6** (1955) 89–91.
Carnap, R., 'On the Character of Philosophical Problems', *Philosophy of Science* **1** (1934) 5–19.
Carnap, R. and Bar-Hillel, Y., 'An Outline of the Theory of Semantic Information', *Massachusetts Institute of Technology, Research Laboratory of Electronics, Technical Report #247*, 1952.
Carnap, R., 'Theory and Predication in Science', *Science* **104** (1946) 502–21.
Cartwright, R. L., 'Ontology and the Theory of Meaning', *Philosophy of Science* **21** (1954) 316–25.
Cassirer, E., 'The Influence of Language upon the Development of Scientific Thought', *Journal of Philosophy* **39** (1942) 309–27.
Cassirer, E., *Substance and Function and Einstein's Theory of Relativity*, La Salle, Ill. 1923.
Cavell, S., 'Must We Mean What We Say?', *Inquiry* **1** (1958) 172–212.

Caws, P., 'A Reappraisal of the Conceptual Scheme of Science', *Philosophy of Science* **24** (1957) 221–34.
Chalmers, G. K., 'The Lodestone and the Understanding of Matter in Seventeenth Century England', *Philosophy of Science* **4** (1937) 75–95.
Chappell, V. C., *The Philosophy of Mind*, Englewood Cliffs, N.J. 1962.
Chihara, C. S., 'Mathematical Discovery and Concept Formation', *Philosophical Review* **72** (1963) 17–34.
Chihara, C. S., *On Mathematical Discovery*, Ph.D. Thesis, University of Washington, 1960.
Chihara, C. S., 'Our Ontological Commitment to Universals', *Nous* **2** (1968) 25–46.
Child, A., 'On the Theory of Categories', *Philosophy and Phenomenological Research* **7** (1946–1947) 316–35.
Child, A., 'Toward a Functional Definition of the A Priori', *Journal of Philosophy* **41** (1944) 155–60.
Chin, Y. L., 'The Principle of Induction and the A Priori', *Journal of Philosophy* **37** (1940) 178–87.
Chisholm, R. M., 'Identity Through Possible Worlds: Some Questions', *Nous* **1** (1967) 1–8.
Chisholm, R. M., 'A Note on Carnap's Meaning Analysis', *Philosophical Studies* **6** (1955) 87–9.
Chomsky, N., *Aspects of the Theory of Syntax*, Cambridge, Mass. 1965.
Chomsky, N., 'Aspects of the Theory of Syntax', *Philosophy and Phenomenological Research* **28** (1967) 278–80.
Chomsky, N., *Cartesian Linguistics*, New York 1966.
Chomsky, N., *Current Issues in Linguistic Theory*, The Hague 1964.
Chomsky, N., *Language and Mind*, New York 1968.
Chomsky, N., 'Quine's Empirical Assumptions', *Synthese* **19** (1969) 53–68.
Chomsky, N., 'Recent Contributions to the Theory of Innate Ideas', *Synthese* **17** (1967) 2–11.
Chomsky, N., *The Logical Structures of Linguistic Theory*, Cambridge, Mass. 1961.
Chomsky, N., 'Three Models for the Description of Language', in *Institute of Radio Engineers, Transactions on Information Theory*, Vol. 1T-2 (1956) pp. 113–24.
Chomsky, N., 'On Certain Formal Properties of Grammars', *Information and Control* **2** (1959) 137–67.
Chomsky, N., 'Recent Contributions to the Theory of Innate Ideas', *Synthese* **17** (1967) 2–11.
Chomsky, N., *Syntactic Structures*, New York 1957.
Chomsky, N., 'Systems of Syntactic Analysis', *Journal of Symbolic Logic* **18** (1953) 242–56.
Chomsky, N., *Topics in the Theory of Generative Grammar*, New York 1966.
Chomsky, N., *Transformational Analysis*, Ph.D. Thesis, University of Pennsylvania, 1960.
Chomsky, N., *A Transformational Approach to Syntax*, Cambridge, Mass. 1959.
Chomsky, N., *The Transformational Basis of Syntax*, Cambridge, Mass. 1959.
Chomsky, N. and Miller, G. A., *Introduction to the Formal Analysis of Natural Language*, Cambridge, Mass. 1962.
Church, A., 'The Need for Abstract Entities in Semantic Analysis', *Daedalus* **80** (1951–1954) 100–12.

GENERAL BIBLIOGRAPHY 255

Churchman, C. W., 'Concepts Without Primitives', *Philosophy of Science* **20** (1953) 257–66.
Clark, R., 'On What is Naturally Necessary', *Journal of Philosophy* **62** (1965) 613–25.
Clifford, W. K., 'On the Space-Theory of Matter', *Cambridge Philosophical Society, Proceedings* **2** (1876) 157–8.
Cohen, F. S., 'The Relativity of Philosophical Systems and the Method of Systematic Relativism', *Journal of Philosophy* **36** (1939) 57–72.
Cohen, L. J., 'A Relation of Counterfactual Conditionals to Statements of What Makes Sense', *Aristotelian Society, Proceedings*, n.s. vol. **55** (1954–1955) 45–82.
Cohen, L. J., 'Are Philosophical Theses Relative to Language?', *Analysis* **9** (1948–1949) 72–7.
Cohen, L. J., *The Diversity of Meaning*, London 1962.
Cole, R., 'Ptolemy and Copernicus', *Philosophical Review* **71** (1962) 476–82.
Collins, J., 'Mr. Lewis and the A Priori', *Journal of Philosophy* **45** (1948) 561–72.
Colodny, R. G. (ed.), *Beyond the Edge of Certainty*, Englewood Cliffs, N.J. 1965.
Colodny, R. G. (ed.), *Frontiers of Science and Philosophy*, Pittsburgh 1962.
Colodny, R. G. (ed.), *Mind and Cosmos*, Pittsburgh 1966.
Cooley, H. T., 'On Mr. Toulmin's Revolution in Logic', *Journal of Philosophy* **56** (1959) 297–319.
Cooley, J. C., 'Professor Goodman's *Fact, Fiction and Forecast*', *Journal of Philosophy* **54** (1957) 293–311.
Copeland, Sr., A. H., 'Mathematical Proof and Experimental Proof', *Philosophy of Science* **33** (1966) 303–16.
Copernicus, N., *De Revolutionibus Orbium Coelestium*, 1543, Repr. Facsimile ed., Johnson Reprints, 1965.
Copi, I. M., 'Gödel and the Synthetic A Priori: A Rejoinder', *Journal of Philosophy* **47** (1950) 633–36.
Copi, I. M., 'Language Analysis and Metaphysical Inquiry', in *The Linguistic Turn* (ed. by R. Rorty), Chicago 1967, pp. 127–31.
Copi, I. M., 'Modern Logic and the Synthetic A Priori', *Journal of Philosophy* **46** (1949) 243–5.
Cornman, J. W., 'Linguistic Frameworks and Metaphysical Questions', *Inquiry* **7** (1964) 129–42.
Cornman, J. W., 'Mental Terms, Theoretical Terms, and Materialism', *Philosophy of Science* **35** (1968) 45–63.
Cornman, J. W., 'Sellars, Scientific Realism and Sense', *Review of Metaphysics* **23** (1970) 417–51.
Costello, H., 'Radical Empiricism and the Concept of "Experienced As"', *Journal of Philosophy* **45** (1948) 225–48.
Cousin, D. R., 'Carnap's Theories of Truth', *Mind* **59** (1950) 1–22.
Craig, E. J., 'Phenomenal Geometry', *British Journal for the Philosophy of Science* **20** (1969) 121–34.
Crawshay-Williams, R., 'Equivocal Confirmation', *Analysis* **11** (1950–1951) 73–80.
Crombie, A. C. (ed.), *Scientific Change*, Symposium on the History of Science, University of Oxford 1961; New York 1963.
Cross, R. C., 'Category Differences', *Aristotelian Society, Proceedings* **59** (1958–1959) 255–70.
Crossley, J. and Dummett, M. (eds.), *Formal Systems and Recursive Functions*, Amsterdam 1965.

Cunningham, G. W., 'Meaning, Reference, and Significance', *Philosophical Review* **47** (1948) 155–75.
d' Abro, A., *The Evolution of Scientific Thought from Newton to Einstein*, 2nd ed., New York 1950.
d' Abro, A., *The Rise of the New Physics*, 2 vols., New York 1951.
Dalkey, N., 'The Limits of Meaning', *Philosophy and Phenomenological Research* **4** (1943–1944) 401–9.
Danto, A. and Morgenbesser, S. (eds.), *Philosophy of Science*, New York 1960.
Davidson, D., 'Emeroses by Other Names', *Journal of Philosophy* **63** (1966) 778–80.
Davidson, D. and Hintikka, J. (eds.), *Words and Objections Essays on the Work of W.V. Quine*, Dordrecht 1969.
Davies, D., 'A Priori Concepts', *Analysis* **2** (1934–1935) 20–4.
Dennis, W. (ed.), *Current Trends of Psychological Theory*, Pittsburgh 1951.
Dingle, H., 'Reason and Experiment in Relation to the Special Theory of Relativity', *British Journal for the Philosophy of Science* **15** (1964–1965) 41–61.
Dingle, H., 'The Scientific Outlook in 1851 and in 1951', *British Journal for the Philosophy of Science* **2** (1951–1952) 85–105.
Dirac, P. A. M., 'The Evolution of the Physicist's Picture of Nature', *Scientific American* **208** (1963) 45–53.
Dirac, P. A. M., 'On the Analogy Between Classical and Quantum Mechanics', *Reviews of Modern Physics* **17** (1945) 195–9.
Dixon, R. M. W., *Linguistic Science and Logic*, The Hague 1963.
Dixon, R. M. W., 'A Logical Statement of Grammatical Theory as Contained in Halliday's "Categories of the Theory of Grammar"', *Language* **39** (1963) 654–68.
Doan, F. M., 'On the Organizational Base of Language with Special Reference to Mathematical Models', *Philosophy and Phenomenological Research* **21** (1960–1961) 239–47.
Dobzhansky, T., 'Scientific Explanation – Chance and Anti-Chance in Organic Evolution', *Delaware Seminar in the Philosophy of Science*, vol. I, 1963, pp. 209–22.
Dodwell, P. C., 'Causes of Behaviour and Explanation in Psychology', *Mind* **69** (1960) 1–13.
Donnellan, K. S., *C. I. Lewis and the Foundations of Necessary Truth*, Ph.D. Thesis, Cornell University, 1961.
Donnellan, K. S., 'Necessity and Criteria', *Journal of Philosophy* **59** (1962) 647–58.
Donnellan, K. S., 'Putting Humpty Dumpty Together Again', *Philosophical Review* **77** (1968) 203–15.
Dotterer, R. H., 'The Operational Test of Meaninglessness', *Monist* **44** (1934) 231–7.
Dretske, F. I., 'Can Events Move?', *Mind* **76** (1967) 479–92.
Dretske, F. I., 'Observational Terms', *Philosophical Review* **73** (1964) 25–42.
Dretske, F. I., 'Reasons and Consequences', *Analysis* **28** (1968) 166–8.
Dretske, F. I., *Seeing and Knowing*, Chicago 1969.
Ducasse, C. J., 'How Does One Discover What a Term Means?', *Philosophical Review* **63** (1954) 88–91.
Ducasse, C. J., 'Verification, Verifiability and Meaningfulness', *Journal of Philosophy* **33** (1936) 230–6.
Duhem, P. M. M., *The Aim and Structure of Physical Theory*, Princeton 1962.
Duhem, P. M. M., *Le Système du Monde: Histoire des Doctrines Cosmologiques de Platon à Copernic*, Paris, 1913–59.
Dummett, M., 'Constructionalism', *Philosophical Review* **66** (1957) 47–65.

Duncan-Jones, A. E., 'Definition of Identity of Structure', *Analysis* **2** (1934-1935) 14-8.
Earle, W., 'The Concept of Existence', *Journal of Philosophy* **57** (1960) 22-3.
Ebersole, F. B., 'On Certain Confusions in the Analytic-Synthetic Distinction', *Journal of Philosophy* **53** (1956) 485-94.
Edel, A., 'Interpretation and the Selection of Categories', *California University Publications in Philosophy* **25** (1950) 57-95.
Edwards, R. B., 'The Truth and Falsity of Definitions', *Philosophy of Science* **33** (1966) 76-9.
Einstein, A., *The Meaning of Relativity*, 3rd ed., Princeton 1950.
Einstein, A. and Infeld, L., *The Evolution of Physics*, New York 1954.
Eiseley, L., *Darwin's Century*, New York 1958.
Ellis, B. D., 'A Comparison of Process and Non-Process Theories in the Physical Sciences', *British Journal for the Philosophy of Science* **8** (1957-1958) 45-56.
Ellis, B. D., 'Derived Measurement, Universal Constants, and the Expression of Numerical Laws', *Delaware Seminar in the Philosophy of Science*, vol. 2, 1963, pp. 371-92.
Ellson, D. G., '"The Scientist" Criterion of True Observation', *Philosophy of Science* **30** (1963) 41-52.
Emig, J. A., Fleming, J. T., and Popp, H. M. (eds.), *Language and Learning*, New York 1966.
Epstein, J., 'Quine's Gambit Accepted', *Journal of Philosophy* **55** (1958) 673-83.
Erickson, R. W., 'The Metaphysics of a Logical Empiricist', *Philosophy of Science* **8** (1941) 320-8.
Erwin, E., *The Concept of Meaninglessness*, Baltimore 1970.
Evans, J. L., 'On Meaning and Verification', *Mind* **62** (1953) 1-19.
Eveling, H. S. and Leith, G. O. M., 'When to Use the Paradigm Case Argument', *Analysis* **18** (1957-1958) 150-2.
Ewing, A. C., 'The Linguistic Theory of A Priori Propositions', *Aristotelian Society, Proceedings*, n.s. vol. 40 (1939-1940) 207-44.
Ewing, A. C., 'Meaninglessness', *Mind* **46** (1937) 347-64.
Ezorsky, G., 'On the Interchangeability of Synonyms', *Philosophy and Phenomenological Research* **19** (1958-1959) 536-8.
Farber, M., 'Relational Categories and the Quest for Unity', *Philosophical Review* **43** (1934) 368-79.
Farre, G. L., 'On the Linguistic Foundations of the Problem of Scientific Discovery', *Journal of Philosophy* **65** (1968) 779-94.
Farrell, B. A., 'Experience', in *The Philosophy of Mind* (ed. by V. C. Chappell), Englewood Cliffs, N.J., 1962, pp. 23-48.
Fearing, F., 'An Examination of the Conceptions of Benjamin Whorf in the Light of Theories of Perception and Cognition', in *Language in Culture* (ed. by H. Hoijer), Chicago, 1954, pp. 47-81.
Feibleman, J. K., 'Mathematics and its Application in the Sciences', *Philosophy of Science* **23** (1956) 204-15.
Feigl, H., 'Confirmability and Confirmation', *Revue Internationale de Philosophie* **5** (1951) 268-79.
Feigl, H., 'De Principiis Non Disputandum...?', in *Philosophical Analysis* (ed. by M. Black), Ithaca, 1950, pp. 119-56.
Feigl, H., 'Existential Hypotheses', *Philosophy of Science* **17** (1950) 35-62.

Feigl, H., 'Logical Reconstruction, Realism, and Pure Semiotic', *Philosophy of Science* **17** (1950) 186–95.
Feigl, H., 'The "Mental" and the "Physical"', *Minnesota Studies in the Philosophy of Science*, vol. II (ed. by H. Feigl, M. Scriven, and G. Maxwell), Minneapolis, 1958, pp. 370–497.
Feigl, H., 'Operationism and Scientific Method', *Psychological Review* **52** (1945) 250–9.
Feigl, H., 'Principles and Problems of Theory Construction in Psychology', in *Current Trends of Psychological Theory* (ed. by W. Dennis), Pittsburgh, 1951, pp. 174–213.
Feigl, H., 'Scientific Method Without Metaphysical Presuppositions', *Philosophical Studies* **5** (1954) 17–29.
Feigl, H., 'Some Major Issues and Developments in the Philosophy of Science of Logical Empiricism', *Minnesota Studies in the Philosophy of Science*, vol. I (ed. by H. Feigl and M. Scriven), Minneapolis, 1956, pp. 3–37.
Feigl, H. and Maxwell, G. (eds.), *Current Issues in the Philosophy of Science*, New York 1961.
Feigl, H. and Sellars, W. (eds.), *Readings in Philosophical Analysis*, New York 1949.
Felch, W. F., 'Are There Synthetic A Priori Truths?', *Journal of Philosophy* **47** (1950) 579–84.
Feuer, L. S., 'Analysis and Scientific Practice', *Analysis* **8** (1947–1948) 28–30.
Feuer, L. S., 'The Paradox of Verifiability', *Philosophy and Phenomenological Research* **12** (1951–1952) 24–41.
Feuer, L. S., 'The Principle of Simplicity', *Philosophy of Science* **24** (1957) 109–21.
Feuer, L. S., 'Sociological Aspects of the Relation Between Language and Philosophy', *Philosophy of Science* **20** (1953) 85–100.
Feyerabend, P. K., 'Against Method: Outline of a Anarchistic Theory of Knowledge', *Analyses of Theories and Methods of Physics and Psychology* (ed. by M. Radner and S. Winokur), Vol. IV of *Minnesota Studies in the Philosophy of Science* (1970), pp. 17–130.
Feyerabend, P. K., 'Attempt at a Realistic Interpretation of Experience', *Aristotelian Society, Proceedings*, n.s. vol. **58** (1958) 143–70.
Feyerabend, P. K., 'Classical Empiricism', in *The Methodological Heritage of Newton* (ed. by R. E. Butts, and J. W. Davis), Toronto, 1969, pp. 150–70.
Feyerabend, P. K., 'Explanation, Reduction and Empiricism', in *Minnesota Studies in the Philosophy of Science*, vol. III (ed. by H. Feigl and G. Maxwell), Minneapolis, 1962, pp. 28–97.
Feyerabend, P. K., 'How to be a Good Empiricist – a Plea for Tolerance in Matters Epistemological', *Delaware Seminar in the Philosophy of Science*, vol. 2, New York 1963, pp. 3–39.
Feyerabend, P. K., 'In Defense of Classical Physics', *Studies in History and Philosophy of Science* **1** (1970) 59–86.
Feyerabend, P. K., *Knowledge without Foundations*, Oberlin 1961.
Feyerabend, P. K., 'Metascience', *Philosophical Review* **70** (1961) 396–405.
Feyerabend, P. K., 'On the "Meaning" of Scientific Terms', *Journal of Philosophy* **62** (1965) 266–74.
Feyerabend, P. K., 'Problems of Microphysics', in *Philosophy of Science Today* (ed. by S. Morgenbesser), New York, 1967, pp. 136–47.
Feyerabend, P. K., 'Science without Experience', *Journal of Philosophy* **66** (1969) 791–94.

Feyerabend, P. K., 'Wittgenstein's *Philosophical Investigations*', *Philosophical Review* **64** (1955) 449–83.
Feyerabend, P. K. and Maxwell, G. (eds.), *Mind, Matter and Method, Essays in Philosophy and Science in Honor of Herbert Feigl*, Minneapolis 1966.
Fiala, S., 'The Experiment and its Role in the Theory of Knowledge', *Philosophy of Science* **18** (1951) 253–8.
Findlay, J. N., 'Some Reflections on Meaning', *The Indian Journal of Philosophy* **1** (1959) 8–16.
Fine, A. I., 'Consistency, Derivability, and Scientific Change', *Journal of Philosophy* **64** (1967) 231–40.
Finkelstein, D., 'Matter, Space and Logic', in *Boston Studies in the Philosophy of Science*, vol. V (ed. by R. Cohen and M. Wartofsky), Dordrecht, 1969, pp. 199–215.
Firth, R., 'Radical Empiricism and Perceptual Relativity', *Philosophical Review* **59** (1950) 164–83, 319–31.
Firth, R., 'Ultimate Evidence', *Journal of Philosophy* **53** (1956) 732–9.
Fisher, A. L. and Murray, G. B. (eds.), *Philosophy and Science as Modes of Knowing: Selected Essays*, New York 1969.
Fisher, M., 'Category-Absurdities', *Philosophy and Phenomenological Research* **24** (1963) 260–7.
Fisk, M., 'Analyticity and Conceptual Revision', *Journal of Philosophy* **63** (1966) 627–37.
Fitch, F. B., 'Actuality, Possibility, and Being', *Review of Metaphysics* **3** (1950) 367–84.
Fitch, F. B., 'Physical Continuity', *Philosophy of Science* **3** (1936) 486–93.
Fitch, F. B., 'Some Logical Aspects of Reference and Existence', *Journal of Philosophy* **57** (1960) 20–1.
Fleming, B. N., 'Recognizing and Seeing As', *Philosophical Review* **66** (1957) 161–79.
Flew, A. G. N. (ed.), *Logic and Language*, 1st and 2nd series, Garden City, N. Y. 1965.
Fodor, J. A., 'Projection and Paraphrase in Semantics', *Analysis* **21** (1960–1961) 73–7.
Fodor, J. A and Katz, J. J., 'The Structure of a Semantic Theory', *Language* **39** (1963) 170–210.
Fodor, J. A., 'What Do You Mean?', *Journal of Philosophy* **57** (1960) 499–507.
Fodor, J. A. and Katz, J. J. (eds.), *The Structure of Language*, Englewood Cliffs, N. J. 1964.
Føllesdal, D., 'Comments on Stenius's 'Mood and Language-Game'', *Synthese* **17** (1967) 275–80.
Fraassen, B. van, 'On the Extension of Beth's Semantics of Physical Theories', *Philosophy of Science* **37** (1970) 325–39.
Frank, P., 'Comments on Realistic versus Phenomenalistic Interpretations', *Philosophy of Science* **17** (1950) 166–9.
Frank, P., 'The Mechanical versus the Mathematical Conception of Nature', *Philosophy of Science* **4** (1937) 41–74.
Frank, P., 'Metaphysical Interpretations of Science', *British Journal for the Philosophy of Science* **1** (1950–1951) 60–91.
Frank, P., *Philosophy of Science*, Englewood Cliffs, N. J. 1957.
Frank, P. (ed.), *The Validation of Scientific Theories*, Boston 1957.
Frankel, A. A., 'Epistemology and Logic', *Logic and Language, Studies Dedicated to Professor Rudolf Carnap* (ed. by B. H. Kazemier and D. Vuysje), Dordrecht, 1962, pp. 6–10.

Freund, J. E., 'On the Confirmation of Scientific Theories,' *Philosophy of Science* **17** (1950) 87–94.
Friedrich, L. W. (ed.), *The Nature of Physical Knowledge*, Bloomington, Indiana 1960.
Fries, H. S., 'On an Empirical Criterion of Meaning', *Philosophy of Science* **3** (1936) 143–51.
Fries, H. S., 'Historical Interpretation and Culture Analysis', *Journal of Philosophy* **49** (1952) 340–50.
Furth, H. G., *Thinking without Language. Psychological Implications of Deafness*, New York 1966.
Gahringer, R. E., 'Can Games Explain Language?', *Journal of Philosophy* **56** (1959) 661–7.
Gahringer, R. E., 'Some Observations on the Distinction between Analytic and Synthetic Propositions', *Journal of Philosophy* **51** (1954) 425–36.
Galilei, Galileo, *Dialogues Concerning the Two Chief Systems of the World* (transl. by Drake and Stillman), Berkeley, California 1962.
Garnett, A. C., 'Arthur Pap's Analysis of Necessary Propositions', *Philosophical Review* **59** (1950) 370–4.
Garnett, A. C., 'Scientific Method and the Concept of Emergence', *Journal of Philosophy* **39** (1942) 477–86.
Geach, P. T., *Mental Acts*, London 1957.
Geach, P. T., 'Quine on Classes and Properties', *Philosophical Review* **62** (1953) 409–12.
Gellner, E., 'Ideal Language and Kinship Structure', *Philosophy of Science* **24** (1957) 235–42.
Gellner, E., *Words and Things*, London 1959.
Gerr, S., 'Language and Science', *Philosophy of Science* **9** (1942) 146–61.
Gewirth, A., 'The Distinction between Analytic and Synthetic Truths', *Journal of Philosophy* **50** (1953) 397–425.
Gledymin, J., 'The Paradox of Meaning Variance', *British Journal for the Philosophy of Science* **21** (1970) 257–68.
Glenn, E. S., 'On the Developmental Theory of Languages', *American Anthropologist* **59** (1957) 537–8.
Gödel, K., 'Zum Intuitionistic Logic', *Ergebnisse Eines Mathematischen Kolloquiums*, Leipzig, 1933, p.40.
Gödel, K., 'Über formal unentscheidbare Sätze der *Principia Mathematica* und verwandter System I,' *Monatshefte für Mathematik und Physik* **38** (1931) 173–98.
Goldstein, L. J., 'The Inadequacy of the Principle of Methodological Individualism', *Journal of Philosophy* **53** (1956) 801–13.
Gombrich, E. H., *Art and Illusion*, New York 1965.
Gomperz, H., 'The Meanings of "Meaning"', *Philosophy of Science* **8** (1941) 157–83.
Goodman, N., *Fact, Fiction and Forecast*, 2nd ed. Indianapolis 1965.
Goodman, N., *Languages of Art: An Approach to a Theory of Symbols*, Indianapolis 1968.
Goodman, N., 'An Improvement in the Theory of Simplicity', *Journal of Symbolic Logic* **13** (1948) 228–9.
Goodman, N., 'New Notes on Simplicity', *Journal of Symbolic Logic* **17** (1952) 189–91.
Goodman, N., 'On Likeness of Meaning', *Analysis* **10** (1948–1950) 1–8.
Goodman, N., 'On Some Differences About Meaning', *Analysis* **13** (1952–1953) 90–7.

Goodman, N., 'Recent Developments in the Theory of Simplicity', *Philosophy and Phenomenological Research* **19** (1958–1959) 429–46.
Goodman, N., 'Safety, Strength, Simplicity', *Philosophy of Science* **28** (1961) 150–1.
Goodman, N., 'Science and Simplicity', in *Philosophy of Science Today* (ed. by S. Morgenbesser), New York 1967, pp. 68–78.
Goodman, N., 'Sense and Certainty', *Philosophical Review* **61** (1952) 160–7.
Goodman, N., 'Some Reflections on the Theory of Systems', *Philosophy and Phenomenological Research* **9** (1948–1949) 620–5.
Goodman, N., *The Structure of Appearance*, Cambridge, Mass. 1951.
Goodman, N., 'Two Replies', *Journal of Philosophy* **64** (1967) 286–7.
Greenberg, J. H., 'Concerning Inferences from Linguistic to Nonlinguistic Data', in *Language in Culture* (ed. by H. Hoijer), Chicago, 1954, pp. 3–19.
Gregory, R. L., 'On Physical Model Explanations in Psychology', *British Journal for the Philosophy of Science* **4** (1953–1954) 192–7.
Grene, M., 'Two Evolutionary Theories', *British Journal for the Philosophy of Science* **9** (1958–1959) 110–27, and 185–93.
Grice, H. P. and Strawson, P. F., 'In Defense of a Dogma', *Philosophical Review* **65** (1956) 141–58.
Grice, H. P., 'Utterer's Meaning, Sentence-Meaning and Word-Meaning', *Foundations of Language* **3** (1968) 225–42.
Gross, L. (ed.), *Symposium on Sociological Theory*, Evanston, Illinois, 1959.
Grünbaum, A., 'The Denial of Absolute Space and the Hypothesis of a Universal Nocturnal Expansion, a Rejoinder to George Schlesinger', *Australasian Journal of Philosophy* **45** (1967) 61–91.
Grünbaum, A., 'The Duhemian Argument', *Philosophy of Science* **27** (1960) 75–87.
Grünbaum, A., 'The Falsifiability of Theories Total or Partial: A Contempory Evaluation of the Duhem-Quine Thesis', *Synthese* **14** (1962) 17–33.
Grünbaum, A., 'The Genesis of the Special Theory of Relativity', in *Current Issues in the Philosophy of Science* (ed. by H. Feigl and G. Maxwell), New York, 1961, pp. 43–53.
Grünbaum, A., *Geometry and Chronometry in Philosophical Perspective*, Minneapolis 1968.
Grünbaum, A., 'Geometry, Chronometry, and Empiricism', *Minnesota Studies in the Philosophy of Science*, vol. III (ed. by H. Feigl and G. Maxwell), Minneapolis, 1962, pp. 405–526
Grünbaum, A., 'Logical and Philosophical Foundations of the Special Theory of Relativity', in *Philosophy of Science* (ed. by A. Danto and S. Morgenbesser), New York, 1960, pp. 399–434.
Grünbaum, A., *Philosophical Problems of Space and Time*, New York 1963.
Grünbaum, A., 'The Relevance of Philosophy to the History of the Special Theory of Relativity', *Journal of Philosophy* **59** (1962) 561–74.
Grünbaum, A., 'Space and Time', in *Philosophy of Science Today* (ed. by S. Morgenbesser), New York, 1967, pp. 125–35.
Grünbaum, A., 'The Special Theory of Relativity as a Case Study of the Importance of the Philosophy of Science for the History of Science', *Delaware Seminar in the Philosophy of Science*, vol. **2**, New York, 1963, pp. 171–204.
Hahn, H., 'Logic, Mathematics and Knowledge of Nature', in *Logical Positivism* (ed. by A. J. Ayer), New York, 1959, pp. 147–61.

Hahn, L. E., 'Of Shoes and Ships and Sealing Wax, and Cabbages and Kings', *Journal of Philosophy* **55** (1958) 45–57.
Haines, G., 'Art Forms and Science Concepts', *Journal of Philosophy* **40** (1943) 482–91.
Hall, A. R., *The Scientific Revolution: 1500-1800*, London 1962.
Hall, A. R. (ed.), *The Making of Modern Science*, Leicester 1960.
Hall, A. R., *The Rise of Modern Science*, New York 1962.
Hall, E. W., 'A Categorial Analysis of Value', *Philosophy of Science* **14** (1947) 333–44.
Hall, E. W., *Categorial Analysis, Selected Essays of Everett W. Hall on Philosophy, Value, Knowledge, and the Mind* (ed. by E. M. Adams), Chapel Hill 1964.
Hall, E. W., 'Is Philosophy a Science?', *Journal of Philosophy* **39** (1942) 113–8.
Hall, E. W., 'Logical Subjects and Physical Objects', *Philosophy and Phenomenological Research* **17** (1956–1957) 478–82.
Hall, E. W., 'The Metaphysics of Logic', *Philosophical Review* **58** (1949) 16–25.
Hall, E. W., 'Of What Use is Metaphysics?', *Journal of Philosophy* **33** (1936) 236–45.
Hall, E. W., 'On the Nature of the Predicate, "Verified"', *Philosophy of Science* **14** (1947) 123–31.
Hall, E. W., *Our Knowledge of Fact and Value*, Chapel Hill 1961.
Hall, E. W., *Philosophical Systems, A Categorical Analysis*, Chicago 1960.
Hall, R. J., 'Assuming: One Set of Positing Words', *Philosophical Review* **67** (1958), 52–75.
Hall, R. J., 'Conceptual Reform – One Task of Philosophy', *Aristotelian Society Proceedings* **61** (1960–1961) 169–88.
Hall, R. J., 'Kuhn and the Copernican Revolution', *British Journal for the Philosophy of Science* **21** (1970) 196–7.
Halldén, S. I., *The Logic of Nonsense*, Uppsala 1949.
Halliday, M. A. K., 'Categories of the Theory of Grammar', *Word* **17** (1961) 241–92.
Hallie, P. P., 'Wittgenstein's Grammatical-Empirical Distinction', *Journal of Philosophy* **60** (1963) 517–28.
Hamlyn, D. W., 'Categories, Formal Concepts and Metaphysics', *Philosophy* **34** (1959) 111–24.
Hamlyn, D. W., *Seeing Things as They Are*, Inaugural Lecture delivered at Birkbeck College, London 1965.
Hampshire, S., 'The Interpretation of Language: Words and Concepts', in *British Philosophy in the Mid-Century* (ed. by C. A. Mace), London, 1957, pp. 267–79.
Hanson, N. R., 'An Anatomy of Discovery', *Journal of Philosophy* **64** (1967) 321–52.
Hanson, N. R., *The Concept of the Positron: A Philosophical Analysis*, Cambridge, England 1963.
Hanson, N. R., 'Discovering the Positron', *British Journal for the Philosophy of Science* **12** (1961–1962) 194–215, 299–314.
Hanson, N. R., 'The Genetic Fallacy Revisited', *American Philosophical Quarterly* **4** (1967) 101–13.
Hanson, N. R., 'More on "The Logic of Discovery"', *Journal of Philosophy* **57** (1960) 182–8.
Hanson, N. R., 'Observation and Interpretation', in *Philosophy of Science Today* (ed. by S. Morgenbesser), New York, 1967, pp. 89–99.
Hanson, N. R., *Patterns of Discovery*, Cambridge, England 1958.
Hanson, N. R., 'Some Philosophical Aspects of Contemporary Cosmologies', *Delaware Seminar in the Philosophy of Science*, Vol. II, New York, 1963, pp. 465–82.

Hanson, N. R., 'The Irrelevance of History of Science to the Philosophy of Science', *Journal of Philosophy* **59** (1962) 574–86.
Harman, G., 'An Introduction to "Translation and Meaning", Chapter Two of *Word and Object*', *Synthese* **19** (1969) 14–26.
Harman, G., 'Psychological Aspects of the Theory of Syntax', *Journal of Philosophy* **64** (1967) 75–87.
Harman, G., 'Sellars' Semantics', *Philosophical Review* **79** (1970) 404–19.
Harrah, D., 'A Model of Communication', *Philosophy of Science* **23** (1956) 333–41.
Harré, R., *The Anticipation of Nature*, London 1965.
Harré, R., *An Introduction to the Logic of the Sciences*, London 1960.
Harré, R., *Matter and Method*, London 1964.
Harré, R., 'Tautologies and the Paradigm-Case Argument', *Analysis* **18** (1957–1958) 94–7.
Harré, R., *Theories and Things*, London 1961.
Harré, R., *The Sciences: Their Origin and Methods*, Glasgow 1967.
Harré, R. (ed.), *Scientific Thought 1900–1960*, Oxford 1969.
Harré, R., *Some Nineteenth Century British Scientists*, New York 1969.
Harrison, B., 'Category Mistakes and Rules of Language', *Mind* **73** (1965) 309–25.
Hartman, R. S., 'The Epistemology of the *A Priori*', *Philosophy and Phenomenological Research* **9** (1948–1949) 731–36.
Hartung, F. E., 'Sociological Foundations of Modern Science', *Philosophy of Science* **14** (1947) 68–95.
Harvey, J. H., 'A Note on Categories', *Journal of Philosophy* **44** (1947) 162–5.
Hatcher, W. S., 'Logical Truth and Logical Implication', *Journal of Symbolic Logic* **31** (1966) 561.
Hawkins, D., *The Language of Nature*, San Francisco 1964.
Hayner, P. C., 'Analogical Predication', *Journal of Philosophy* **55** (1958) 855–62.
Heelan, P. A., 'Scientific Objectivity and Framework Transpositions', *Philosophical Studies in Ireland* **19** (1970) 55–70.
Heijenoort, J. van, *From Frege to Gödel*, Cambridge, Mass. 1967.
Heisenberg, W., 'The Development of the Interpretation of the Quantum Theory', in *Neils Bohr and the Development of Physics* (ed. by W. Pauli), New York, 1955, pp. 12–29.
Heisenberg, W., *Physics and Philosophy*, New York 1958.
Helmer, O., 'The Significance of Undecidable Sentences', *Journal of Philosophy* **34** (1937) 490–4.
Hempel, C. G., *Aspects of Scientific Explanation*, New York 1965.
Hempel, C. G., 'Explanation in Science and History', in *Frontiers of Science and Philosophy* (ed. by R. G. Colodny), Pittsburgh, 1962, pp. 7–33.
Hempel, C. G., 'The Function of General Laws in History', *Journal of Philosophy* **39** (1942) 35–48.
Hempel, C. G., *Fundamentals of Concept Formation in Empirical Science*, Chicago 1952.
Hempel, C. G., 'General System Theory and the Unity of Science', *Human Biology* **23** (1951) 313–22.
Hempel, C. G., 'Geometry and Empirical Science', *The American Mathematical Monthly* **52** (1945) 7–17.
Hempel, C. G., 'Implications of Carnap's Work for the Philosophy of Science', in *The Philosophy of Rudolf Carnap* (ed. by P. A. Schilpp), La Salle, Illinois, 1963, pp. 685–710.

Hempel, C. G., 'Introduction to Problems of Taxonomy', in *Field Studies in the Mental Disorders* (ed. by J. Zubin), New York, 1961, pp. 3–23.
Hempel, C. G., 'A Logical Appraisal of Operationism', *Scientific Monthly* **79** (1954) 215–20.
Hempel, C. G., 'A Note on Semantic Realism', *Philosophy of Science* **17** (1950) 169–73.
Hempel, C. G., 'Empirical Statements and Falsifiability', *Philosophy* **33** (1958) 342–8.
Hempel, C. G., 'On the Nature of Mathematical Truth', *The American Mathematical Monthly* **52** (1945) 543–56.
Hempel, C. G., 'Problems and Changes in the Empiricist Criterion of Meaning', *Revue Internationale de Philosophie* **4** (1950) 41–63.
Hempel, C. G., 'Reflections on Nelson Goodman's *The Structure of Appearance*', *Philosophical Review* **62** (1953) 108–16.
Hempel, C. G., 'Scientific Explanation', in *Philosophy of Science Today* (ed. by S. Morgenbesser), New York, 1967, pp. 79–88.
Hempel, C. G., 'The Theoretician's Dilemma', *Minnesota Studies in the Philosophy of Science*, vol. II (ed. by H. Feigl, M. Scriven and G. Maxwell), Minneapolis, 1958, pp. 37–98.
Hems, J. M., 'Learning the Language', *Philosophy and Phenomenological Research* **26** (1965–1966) 561–77.
Henkin, L., 'Some Notes on Nominalism', *Journal of Symbolic Logic* **18** (1953) 19–29.
Henkin, L., 'Two Concepts from the Theory of Models', *Journal of Symbolic Logic* **21** (1956) 28–32.
Henkin, L., Suppes, P., and Tarski, A. (eds.), *The Axiomatic Method with Special Reference to Geometry and Physics*, Proceedings of an International Symposium held at the University of California, Berkeley, December 26, 1957 – January 4, 1958, Amsterdam 1959.
Henle, P., 'On the Certainty of Empirical Statements', *Journal of Philosophy* **44** (1947) 625–32.
Henle, P., 'The Status of Emergence', *Journal of Philosophy* **39** (1942) 486–93.
Herburt, G. K., 'The Analytic and the Synthetic', *Philosophy of Science* **26** (1959) 104–13.
Hervey, H., 'The Problem of the Model Language Game in Wittgenstein's Later Philosophy', *Philosophy* **37** (1961) 333–51.
Herivel, J. W., 'Aspects of French Theoretical Physics in the Nineteenth Century', *British Journal for the History of Science* **3** (1966–1967) 109–32.
Herzberger, H. G., 'Logical Consistency of Language', in *Language and Learning* (ed. by J. A. Emig, J. T. Fleming, and H. M. Popp), New York, 1966, pp. 250–63.
Herzberger, H. G., 'The Truth-Conditional Consistency of Natural Language', *Journal of Philosophy* **64** (1967) 29–35.
Hesse, M. B., 'Analogy and Confirmation Theory', *Philosophy of Science* **31** (1964) 319–27.
Hesse, M. B., 'Fine's Criteria of Meaning Change', *Journal of Philosophy* **65** (1968) 46–52.
Hesse, M. B., 'Models and Analogies in Science', *Newman History and Philosophy of Science Series #14*, London 1963.
Hesse, M. B., 'Models in Physics', *British Journal for the Philosophy of Science* **4** (1953–1954) 198–214.
Hesse, M. B., 'Operational Definition and Analogy in Physical Theories', *British Journal for the Philosophy of Science* **2** (1951–1952) 281–94.

Hesse, M. B., 'Theories, Dictionaries, and Observation', *British Journal for the Philosophy of Science* **9** (1958–1959) 12–28.
Hillman, D. J., 'On Grammars and Category Mistakes', *Mind* **72** (1963) 224–34.
Hinshaw, V., 'The Given', *Philosophy and Phenomenological Research* **18** (1957–1958) 312–25.
Hintikka, J., 'A Program and a Set of Concepts for Philosophical Logic', *Monist* **51** (1967) 69–92.
Hintikka, J., 'Are Logical Truths Analytic?', *Philosophical Review* **74** (1965) 178–203.
Hintikka, J., 'Are Mathematical Truths Synthetic A Priori?', *Journal of Philosophy* **65** (1968) 640–51.
Hintikka, J., 'Behavioural Criteria of Radical Translation', *Synthese* **19** (1969) 69–81.
Hintikka, J., 'Individuals, Possible Worlds, and Epistemic Logic', *Nous* **1** (1967) 33–62.
Hintikka, J., 'Kant on Mathematical Method', *Monist* **51** (1967) 352–75.
Hintikka, J., 'Towards a Theory of Inductive Generalization', in *Logic, Methodology and Philosophy of Science*, Vol. II (ed. by Y. Bar-Hill), Amsterdam, 1964, pp. 274–88.
Hintikka, J., *The Philosophy of Mathematics*, London 1969.
Hiz, H., 'Methodological Aspects of the Theory of Syntax', *Journal of Philosophy* **64** (1967) 67–74.
Hockett, C. F., 'Chinese versus English: An Explanation of the Whorfian Theses', in *Language in Culture* (ed. by H. Hoijer), Chicago, 1954, pp. 106–23.
Hofstadter, A., 'The Question of Categories', *Journal of Philosophy* **48** (1951) 173–85.
Hofstadter, A., 'On Semantic Problems', *Journal of Philosophy* **35** (1938) 225–32.
Hoijer, H. (ed.), *Language in Culture*, Chicago 1954.
Hoijer, H., 'The Sapir-Whorf Hypothesis', in *Language in Culture* (ed. by H. Hoijer), Chicago, 1954, pp. 92–105.
Holton, G., 'On the Origins of the Special Theory of Relativity', *American Journal of Physics* **28** (1960) 627–36.
Hooker, C. (ed.), *Contemporary Research in the Foundations of Quantum Theory*, Dordrecht 1973.
Hooker, C. A., 'Goodman, "Grue" and Hempel', *Philosophy of Science* **35** (1968) 232–48.
Hoyle, F., *The Nature of the Universe*, New York 1960.
Humphries, B. M., 'Indeterminacy of Translation and Theory', *Journal of Philosophy* **67** (1970) 167–78.
Humphreys, W. C., *Anomalies and Scientific Theories*, San Francisco 1968.
Humphreys, W. C., 'Galileo, Falling Bodies and Inclined Places', *British Journal for the History of Science* **3** (1966–1967) 225–44.
Hutten, E. H., 'The Role of Models in Physics', *British Journal for the Philosophy of Science* **4** (1953–1954) 281–301.
Inhelder, B. and Piaget, J., *Early Growth of Logic in the Child*, New York 1964.
Inhelder, B. and Piaget, J., *The Child's Conception of Space*, London 1956.
Inhelder, B. and Piaget, J., *Growth of Logical Thinking from Childhood to Adolescence*, New York 1958.
Jammer, M., *Concepts of Mass in Classical and Modern Physics*, Cambridge, Mass. 1961.
Jammer, M., *Concepts of Space; The History of Theories of Space in Physics*, Cambridge, Mass, 1954.
Jammer, M., *The Conceptual Development of Quantum Mechanics*, New York 1960.
Jeffreys, H., *Operational Methods in Mathematical Physics*, Cambridge, England 1927.

Jeffreys, H., *Scientific Inference*, Cambridge, England 1931.
Jobé, E. K., 'Some Recent Work on the Problem of Law', *Philosophy of Science* **34** (1967) 363–81.
Jonas, H., 'The Scientific and Technological Revolutions', *Philosophy Today* **15** (1971) 76–101.
Kahane, H., 'Nelson Goodman's Entrenchment Theory', *Philosophy of Science* **32** (1965) 377–83.
Kaminsky, J., 'Ontology and Language', *Philosophy and Phenomenological Research* **23** (1962–1963) 176–91.
Kantor, J. P., 'The Role of Language in Logic and Science', *Journal of Philosophy* **35** (1938) 449–63.
Kaplan, A., 'Content Analysis and the Theory of Signs', *Philosophy of Science* **10** (1943) 230–47.
Kaplan, A., 'Definition and Specification of Meaning', *Journal of Philosophy* **43** (1946) 281–8.
Kattsoff, L. O., 'Observation and Interpretation in Science', *Philosophical Review* **56** (1947) 682–9.
Katz, J. J., *Philosophy of Language*, New York 1966.
Katz, J. J., 'Recent Issues in Semantic Theory', *Foundations of Language* **3** (1967) 124–94.
Katz, J. J., 'The Relevance of Linguistics to Philosophy', *Journal of Philosophy* **62** (1965) 590–602.
Katz, J. J., 'A Reply to "Projection and Paraphrase in Semantics"', *Analysis* **22** (1961–1962) 36–41.
Katz, J. J., 'Some Remarks on Quine on Analyticity', *Journal of Philosophy* **64** (1967) 36–52.
Kazemier, B. G. and Vuysje, D. (eds.), *Logic and Language, Studies, Dedicated to Professor Rudolf Carnap on the Occasion of his Seventieth Birthday*, Dordrecht 1962.
Kaufman, A. S., 'The Analytic and the Synthetic', *Philosophical Review* **62** (1953) 421–6.
Kaufman, F., 'The Logical Rules of Scientific Procedure', *Philosophy and Phenomenological Research* **2** (1941–1942) 457–71.
Kaufmann, F., 'Verification, Meaning, and Truth', *Philosophy and Phenomenological Research* **4** (1943–1944) 267–83.
Keene, G. B., 'Analytic Statements and Mathematical Truth', *Analysis* **16** (1955–1956) 86–91.
Keifer, H. and Munitz, M. (eds.), *Language, Belief and Metaphysics*, Albany 1970.
Kemeny, J. G., 'Models of Logical Systems', *Journal of Symbolic Logic* **13** (1948) 16–30
Kemeny, J. G., 'A New Approach to Semantics', parts I and II, *Journal of Symbolic Logic* **21** (1956) 1–27, 149–61.
Kent, W., 'Scientific Naming', *Philosophy of Science* **25** (1958) 185–93.
Keswani, G. H., 'Origin and Concept of Relativity', *British Journal for the Philosophy of Science* **15** (1964–1965) 286–307, **16** (1965–1966) 19–33.
Kiteley, M. J. and Madden, E. H., 'Postulates and Meaning', *Philosophy of Science* **29** (1962) 66–77.
Kleinor, S. A., 'Erotetic Logic and the Structure of Scientific Revolution', *British Journal for the Philosophy of Science* **21** (1970) 149–65.
Klemke, E. D., 'The Laws of Logic', *Philosophy of Science* **33** (1966) 271–7.

Kneale, M. and Kneale, W. C., *The Development of Logic*, Oxford 1962.
Kneale, W. C., 'Are Necessary Truths True by Convention?', *Aristotelian Society, Proceedings*, supplementary **21** (1947) 118–33.
Kneale, W. C., 'The Idea of Invention', *British Academy, Proceedings* **41** (1955) 85–108.
Kneale, W. C., 'Scientific Revolution For Ever', *British Journal for the Philosophy of Science* **19** (1968) 27–42.
Kneale, W. C., 'The Truths of Logic', *Aristotelian Society, Proceedings*, n.s. **46** (1945–1946) 207–34.
Kneale, W. C., 'Universality and Necessity', *British Journal for the Philosophy of Science* **12** (1961–1962) 89–102.
Koertge, N., *A Study of Relations between Scientific Theories: A Test of the General Correspondence Principle* (unpublished Ph.D. dissertation, University of London) 1969.
Koertge, N., 'Inter-Theoretic Criticism and the Growth of Science', in *Boston Studies in the Philosophy of Science*, vol. VIII. (ed. by R. C. Buck and R. S. Cohen), Dordrecht 1971.
Koninck, C. de, 'Concept, Process, and Reality', *Philosophical and Phenomenological Research* **9** (1948–1949) 440–7.
Kordig, C. R., 'Feyerabend and Radical Meaning Variance', *Nous* **4** (1970) 399–404.
Kordig, Carl R., 'Scientific Transitions, Meaning Invariance, and Derivability', *The Southern Journal of Philosophy* **9** (1971) 119–25.
Körner, S., *Categorial Frameworks*, Oxford 1970.
Körner, S., 'Categorial Change and Philosophical Argument', *Proceedings of the Israel Academy of Sciences and Humanities* **3** (No. 10) Jerusalem 1969.
Körner, S. (ed.), 'Colston Research Society', *Observation and Interpretation, a Symposium of Philosophers and Physicists*, New York 1957.
Körner, S., *Conceptual Thinking*, New York 1959.
Körner, S., 'Entailment and the Meaning of Words', *Analysis* **10** (1949–1950) 88–93.
Körner, S., *Experience and Theory: an Essay in the Philosophy of Science*, New York 1966.
Körner, S., *The Philosophy of Mathematics*, London 1960.
Körner, S., 'Reference, Vagueness, and Necessity', *Philosophical Review* **66** (1957)363–76.
Körner, S., 'Some Relations between Philosophical and Scientific Theories', *British Journal for the Philosophy of Science* **17** (1966–1967) 265–79.
Koyré, A., 'Galileo and the Scientific Revolution of the Seventeenth Century', *Philosophical Review* **52** (1943) 333–48.
Kripke, S. A., 'Semantical Analysis of Intuitionistic Logic', *Formal Systems and Recursive Functions* (ed. by J. Crossley and M. Dummett), Amsterdam, 1965, pp. 92–130.
Kripke, S. A., 'Semantical Analysis of Modal Logic I, Normal Modal Propositional Calculi', *Zeitschrift für Mathematische Logik und Grundlagen der Mathematik* **9** (1963) 67–96.
Kripke, A. K., 'Semantical Analysis of Modal Logic' (abstract), *The Journal of Symbolic Logic* **24** (1959) 323–4.
Kripke, S. A., 'Semantical Considerations on Modal and Intuitionistic Logic', *Acta Philosophica Fennica* **16** (1963) 83–94.
Kripke, S. A. and Pour-El, M. B., 'Deduction-Preserving Recursive Isomorphisms Between Theories', *American Mathematical Society Bulletin* **73** (1967) 145–8.
Kuhn, T. S., *The Copernican Revolution*, Cambridge, Mass. 1957.

Kuhn, T. S., 'The Function of Measurement in Modern Physical Science', *Isis* **52** (1961) 161–93.
Kuhn, T. S., 'The Historical Structure of Scientific Discovery', *Science* **136** (1962) 760–4.
Kuhn, T. S., *The Structure of Scientific Revolutions*, Chicago, Ill. 1962.
Kumar, D., 'Logic and Inexact Predicates', *British Journal for the Philosophy of Science* **18** (1967–1968) 211–22.
Kyburg, H. E., 'Two World Views', *Nous* **4** (1970) 337–48.
Kyburg, H. and Smokler, H. (eds.), *Studies in Subjective Probability*, New York 1964.
Lacey, A. R., 'Necessary Statements and Entailment', *Analysis* **22** (1961–1962) 101–6.
LaFleur, L. J., 'Conceptual Relativity', *Journal of Philosophy* **37** (1940) 421–31.
Lakatos, I., 'Proofs and Refutations', *British Journal for the Philosophy of Science* **14** (1963–64) 1–25, 120–39, 221–43, 296–342.
Lakatos, I. and Musgrave, A. (eds.), *Problems in the Philosophy of Science*, Amsterdam 1968.
Lake, B., 'Necessary and Contingent Statements', *Analysis* **12** (1951–1952) 115–22.
Lambert, K., *The Logical Way of Doing Things*, New Haven 1969.
Lambert, K. and Meyer, R. K., 'Universally Free Logic and Standard Quantification Theory', *Journal of Symbolic Logic* **33** (1968) 8–26.
Langford, C. H., 'A Proof that Synthetic A Priori Propositions Exist', *Journal of Philosophy* **46** (1949) 20–4.
Larson, S., 'Analyticity and Impropriety', *Journal of Philosophy* **63** (1966) 640–2.
Lazerowitz, M., 'Are Self-Contradictory Expressions Meaningless?', *Philosophical Review* **58** (1949) 563–84.
Lazerowitz, M., 'The Principle of Verifiability', *Mind* **46** (1937) 372–8.
Lazerowitz, M., 'Strong and Weak Verification', *Mind* **48** (1939) 202–13.
Lean, M., *Sense-Perception and Matter*, London 1953.
LeBlanc, H., 'Confirmation of Laws', *Philosophy of Science* **26** (1959) 364–6.
LeBlanc, H., 'On Logically False Evidence Statements', *Journal of Symbolic Logic* **22** (1957) 345–9.
LeBlanc, H., 'A Revised Version of Goodman's Paradox on Confirmation', *Philosophical Studies* **14** (1963) 49–51.
Lee, D. D., 'Conceptual Implications of an Indian Language', *Philosophy of Science* **5** (1938) 89–102.
Lee, D. S., 'The Construction of Empirical Concepts', *Philosophy and Phenomenological Research* **27** (1966–1967) 183–98.
Leith, G. O. M. and Eveling, H. S., 'When to Use the Paradigm Case Argument', *Analysis* **18** (1957–1958) 150–2.
Lenneberg, E. H., 'Cognition Ethnolinguistics', *Language* **29** (1953) 463–71.
Lenneberg, E. H., 'The Relationship of Language to the Formation of Concepts', *Synthese* **14** (1962) 103–9.
Lepan, A., 'Incompatibilities and Conflicts: Breakdown', *Philosophy of Science* **14** (1947) 261–5.
Leplin, J., 'Meaning Variance and the Comparability of Theories', *British Journal for the Philosophy of Science* **20** (1969) 69–74.
Levi, I., 'Confirmation, Linguistic Invariance, and Conceptual Innovation', *Synthese* **20** (1969) 48–55.
Levi, I., 'Utility and Acceptance of Hypotheses', in *Philosophy of Science Today* (ed. by S. Morgenbesser), New York, 1967, pp. 115–24.

Levin, M. E., 'Fine, Mathematics, and Theory Change', *Journal of Philosophy* **65** (1968) 52-6.
Levy, E., 'On the Problem of the Possibility of a Perceptual World-in-Common', *Philosophy and Phenomenological Research* **28** (1967-1968) 48-57.
Lewis, C. I., *An Analysis of Knowledge and Valuation*, La Salle, Illinois 1946.
Lewis, C. I., 'Experience and Meaning', *Philosophical Review* **43** (1934) 125-46.
Lewis, C. I., 'The Given Element in Empirical Knowledge', *Philosophical Review* **61** (1952) 168-75.
Lewis, C. I., *Mind and the World Order*, New York 1929.
Lindsay, R. B., 'A Critique of Operationalism in Physics', *Philosophy of Science* **4** (1937) 456-70.
Lindsay, R. B. and Margenau, H., *Foundations of Physics*, New York 1957.
Lindsay, R. B., 'The Meaning of Simplicity in Physics', *Philosophy of Science* **4** (1937) 151-67.
Linsky, L., *Semantics and the Philosophy of Language*, Urbana, Illinois 1956.
Linsky, L., 'Some Notes on Carnap's Concept of Intensional Isomorphism and the Paradox of Analysis', *Philosophy of Science* **16** (1949) 343-6.
Liu, S., 'On the Analytic and the Synthetic', *Philosophical Review* **65** (1956) 218-28.
Löb, M. H., 'Extensional Interpretations of Modal Logics', *Journal of Symbolic Logic* **31** (1966) 23-45.
Loewenberg, J., 'What is Empirical?', *Journal of Philosophy* **37** (1940) 281-9.
Lomasky, L., 'Nominalism, Replication and Nelson Goodman', *Analysis* **29** (1968-1969) 156-61.
Lowe, V., 'Categorial Analysis, Metaphysics, and C. I. Lewis', *Journal of Philosophy* **55** (1958) 683-90.
MacCormac, E. R., 'Meaning Variance and Metaphor', *British Journal for the Philosophy of Science* **22** (1971) 145-59.
MacKay, D. M., *Information, Mechanism, and Meaning*, Cambridge, Mass. 1969.
MacKay, D. M., 'The Epistemological Problem for Automata', *Automata Studies* (Annals of Mathematics Studies, **34**), Princeton 1956.
MacKay, D. M., 'What is Cybernetics?', *Smithsonian Institution, Annual Report*, Washington, 1963, pp. 401-7.
McKinney, J. P., 'Concepts and Meanings: A Footnote to Philosophy', *Journal of Philosophy* **52** (1955) 515-8.
McKinney, J. P., 'Philosophical Implications of the Modern Revolution of Thought', *Philosophy and Phenomenological Research* **18** (1957-1958) 35-47.
Mackinnon, E., 'The Role of Conceptual and Linguistic Frameworks', *Proceedings of the Catholic Philosophical Association* **43** (1969) 24-43.
McCarthy, J. and Shannon, C. E. (ed.), *Automata Studies* (Annals of Mathematics Studies, **34**), Princeton 1956.
McKinsey, J. C. C. and Suppes, P., 'On the Notion of Invariance in Classical Mechanics', *British Journal for the Philosophy of Science* **5** (1954-1955) 290-302.
McNaughton, R., 'Axiomatic Systems, Conceptual Schemes, and the Consistency of Mathematical Theories', *Philosophy of Science* **21** (1954) 44-53.
McNaughton, R., 'Conceptual Schemes in Set Theory', *Philosophical Review* **66** (1957) 66-80.
McQuown, N. A., 'Analysis of the Cultural Content of Language Materials', in *Language in Culture* (ed. by H. Hoijer), Chicago, 1954, pp. 20-31.
Madden, E. H. (ed.), *The Structure of Scientific Thought*, Boston 1960.

Malinowski, B., 'The Problems of Meaning in Primitive Language', in *Meaning of Meaning* (ed. by C. Ogden, and I. A. Richards), 10th ed., London, 1949, pp. 297–336.
Malisoff, W. M., 'Emergence Without Mystery', *Philosophy of Science* **6** (1939) 17–24.
Malisoff, W. M., 'What is an Atom?', *Philosophy of Science* **6** (1939) 261–5.
Malisoff, W. M., 'On the Postulates of Empiricism', *Philosophy of Science* **8** (1941) 467–85.
Margenau, H., 'Critical Points in Modern Physical Theory', *Philosophy of Science* **4** (1937) 337–70.
Margenau, H., 'Methodology of Modern Physics', *Philosophy of Science* **2** (1935) 39–48, 169–87.
Margenau, H., 'Philosophical Problems Concerning the Meaning of Measurement in Physics', *Philosophy of Science* **25** (1958) 23–33.
Margenau, H. and Mould, R. A., 'Relativity: An Epistemological Appraisal', *Philosophy of Science* **24** (1957) 297–307.
Margolis, J., 'Feyerabend on Meaning', *Personalist* **51** (1970) 514–21.
Marhenke, P., 'The Criterion of Significance', *American Philosophical Association, Proceedings and Addresses* **23** (1950) 1–21.
Martin, Michael, 'Referential Variance and Scientific Objectivity', *British Journal for the Philosophy of Science* **22** (1971) 17–26.
Martin, R. M., *Belief, Existence and Meaning*, New York 1969.
Martin, R. M., 'Category-Words and Linguistic Frameworks', *Kant-Studien* **54** (1963) 176–80.
Martin, R. M., *The Notion of Analytic Truth*, Philadelphia 1959.
Martin, R. M., 'On Theoretical Constructs and Ramsey Constants', *Philosophy of Science* **33** (1966) 1–13.
Mates, B., 'Analytic Sentences', *Philosophical Review* **60** (1951) 525–34.
Mates, B., 'On the Verification of Statements about Ordinary Language', *Inquiry* **1** (1958) 161–71.
Maxwell, G., 'The Necessary and the Contingent', *Minnesota Studies in the Philosophy of Science*, vol. III (ed. by H. Fiegl, G. Maxwell), Minneapolis, 1962, pp. 398–404.
Maxwell, G., 'Theories, Frameworks, and Ontology', *Philosophy of Science* **29** (1962) 132–8.
Maxwell, N., 'Physics and Common Sense', *British Journal for the Philosophy of Science* **16** (1965–1966) 295–312.
Mayo, B., 'Objects, Events, and Complementarity', *Philosophical Review* **70** (1961) 340–61.
Meckler, L., 'On Goodman's Refutation of Synonymy', *Analysis* **14** (1953–1954) 68–79.
Meehl, P. E. and Sellars, W., 'The Concept of Emergence', *Minnesota Studies in the Philosophy of Science*, vol. I (ed. by H. Feigl and M. Scriven), Minneapolis, 1956, pp. 239–52.
Meehl, P. E. and MacCorquodale, K., 'On a Distinction between Hypothetical Constructs and Intervening Variables', *Psychological Review* **55** (1948) 95–107.
Mehlberg, H., *The Reach of Science*, Toronto 1958.
Mei, T., 'The Logic of Depth Grammar', *Philosophy and Phenomenological Research* **24** (1963–1964) 97–105.
Mellor, W. W., 'Believing the Meaningless', *Analysis* **15** (1954–1955) 41–4.
Mendelbaum, D. (ed.), *Selected Writings in Language, Culture and Personality*, Los Angeles 1949.

Mendelson, W. E. K., 'Physical Models and Physiological Concepts: Explanation in Nineteenth Century Biology', *British Journal for the History of Science* **2** (1964–1965) 201–19.
Menger, K., 'On Variables in Mathematics and in Natural Science', *British Journal for the Philosophy of Science* **5** (1954) 134–42.
Mesthene, E. G., 'On the Status of the Laws of Logic', *Philosophy and Phenomenological Research* **10** (1949–1950) 354–72.
Metcalf, W. V., 'The Reality of the Unobservable', *Philosophy of Science* **7** (1940) 337–41.
Meyer, H., 'On Heuristic Value of Scientific Models', *Philosophy of Science* **17** (1950) 111–23.
Michalos, Alex C., 'Theory of Appraisal of the Growth of Scientific Knowledge', *Studies in History and Philosophy of Science* **1**, (1971) 353–362.
Miller, D. L., 'The A Priori in Contemporary Thought', *Philosophy of Science* **8** (1941) 20–5.
Miller, D. L., 'The Effect of the Concept of Evolution on Scientific Methodology', *Philosophy of Science* **15** (1948) 52–60.
Miller, D. L., 'Meaning and Verification', *Philosophical Review* **52** (1943) 604–9.
Miller, D. L., 'The Nature of the Physical Objects', *Journal of Philosophy* **44** (1947) 352–9.
Miller, D. L., 'Prescriptive Categories in Modern Science', *Journal of Philosophy* **40** (1943) 411–4.
Miller, G. A. and Smith, F. (eds.), *The Genesis of Language: A Psycholinguistic Approach*, Cambridge, Mass. 1966.
Miller, H., 'The Relations of Physics and Biology to Epistemology', *Journal of Philosophy* **32** (1935) 628–40.
Mink, L. O., 'Comment on Stephen Toulmin's "Conceptual Revolutions in Science"', *Synthese* **17** (1967) 92–9.
Montague, R., 'Theories Incomparable with Respect to Relative Interpretability', *Journal of Symbolic Logic* **27** (1962) 195–211.
Moravcsik, J. M. E., 'The Analytic and the Nonempirical', *Journal of Philosophy* **62** (1965) 415–29.
Moravcsik, J. M. E. (ed.), *Aristotle, a Collection of Critical Essays*, Garden City, N.Y. 1967.
Moravcsik, J. M. E. (ed.), 'Aristotle's Theory of Categories', *Aristotle, a Collection of Critical Essays*, (1967) pp. 125–45.
Morgenbesser, S. (ed.), *Philosophy of Science Today*, New York 1967.
Morgenbesser, S., 'The Explanatory-Predicative Approach to Science', *Delaware Seminar in the Philosophy of Science* **1** (1963) 41–56.
Mostowski, A., 'On Direct Products of Theories', *Journal of Symbolic Logic* **17** (1952) 1–31.
Munitz, M. K., 'Kantian Dialectic and Modern Scientific Dialectic', *Journal of Philosophy* **48** (1951) 325–38.
Myers, G. E., 'Inexplicable Analogies', *Philosophy and Phenomenological Research* **22** (1961–1962) 326–33.
Naess, A., 'Interpretation and Preciseness', *Norske Videnskaps-Akademi, Oslo. Historisk-Filosofisk Klasse. Skrifter*, 1953, no. 1.
Nagel, E., 'The Formation of Modern Conceptions of Formal Logic in the Development of Geometry', *Osiris* **7** (1939) 142–224.

Nagel, E., *Logic without Metaphysics and Other Essays in the Philosophy of Science*, Glencoe, Ill. 1957.
Nagel, E., 'Mechanistic Explanation and Organismic Biology', *Philosophy and Phenomenological Research* **11** (1950-1951) 327-38.
Nagel, E., 'The Nature and Aim of Science', *Philosophy of Science Today* (ed. by S. Morgenbesser), New York, 1967, pp. 3-13.
Nagel, E., *On the Logic of Measurement*, Ph.D. Thesis, Columbia 1931.
Nagel, E., 'Operational Analysis as an Instrument for the Critique of Linguistic Signs', *Journal of Philosophy* **39** (1942) 177-89.
Nagel, E., 'Science and Semantic Realism', *Philosophy of Science* **17** (1950) 174-81.
Nagel, E., *Sovereign Reason*, Glencoe, Ill. 1954.
Nagel, E., *The Structure of Science*, New York 1961.
Nagel, E., 'Verifiability, Truth and Verification', *Journal of Philosophy* **31** (1934) 141-8.
Nagel, E., Suppes, P., and Tarski, A. (eds.), *International Congress of Logic, Methodology and Philosophy of Science*, Stanford, Calif. 1960, Proceedings, Stanford 1962.
Nelson, Alvin F., 'The HD Method of Scientific Change', *Southwestern Journal of Philosophy* **2** (1971) 83-92.
Nelson, E. J., 'Categorial Interpretation of Experience', *Philosophy and Phenomenological Research* **13** (1952-1953) 84-95.
Nelson, E. J., 'The Relation of Logic to Metaphysics', *Philosophical Review* **58** (1949) 1-15.
Nelson, E. J., 'The Verification Theory of Meaning', *Philosophical Review* **63** (1954) 182-92.
Nerlich, G. C., 'Sameness, Difference and Continuity', *Analysis* **18** (1957-1958) 144-50.
Newman, S., 'Semantic Problems in Grammatical Systems and Lexemes: A Search for Method', *Language in Culture* (ed. by H. Hoijer) Chicago, 1954, pp. 82-91.
Newton, I., *Isaac Newton's Papers and Letters on Natural Philosophy, and Related Documents* (ed. by I. B. Cohen), Cambridge, Mass. 1958.
Nicod, J., *Foundations of Geometry and Induction*, London 1930.
Nidditch, P. H., 'A Note on Logic and Linguistic Ambiguities', *Analysis* **12** (1951-1952) 115-22.
Novak, M., 'Toward Understanding Aristotle's Categories', *Philosophy and Phenomenological Research* **26** (1965-1966) 117-23.
O'Connor, D. J., 'The Identity of Indiscernibles', *Analysis* **14** (1953-1954) 103-11.
Ogden, C. K. and Richards, I. A. (eds.), *The Meaning of Meaning*, 10th ed., London 1949.
Olds, M. E., 'Ostension and Analyticity', *Philosophy and Phenomenological Research* **18** (1957-1958) 359-67.
Oliver, W. D., 'Logic and Necessity', *Journal of Philosophy* **47** (1950) 69-73.
Olschki, L., 'Galileo's Philosophy of Science', *Philosophical Review* **52** (1943) 349-65.
O'Hair, S. G., 'Putnam on Reds and Greens', *Philosophical Review* **78** (1969) 504-6.
Oppenheim, P. and Putnam, H., 'Unity of Science as a Working Hypothesis', in *Minnesota Studies in the Philosophy of Science*, vol. II (ed. by H. Feigl, M. Scriven and G. Maxwell), Minneapolis, Minnesota, 1958, pp. 3-36.
Palter, R., 'Philosophical Principles and Scientific Theory', *Philosophy of Science* **23** (1956) 111-35.
Pap, A., *The A Priori in Physical Theory*, New York 1946.

Pap, A., *Analytische Erkenntnistheorie, Kritische Übersicht über die Neueste Entwicklung in U.S.A. und England*, Wien 1958.
Pap, A., 'Are All Necessary Propositions Analytic?', *Philosophical Review* **58** (1949) 299–320.
Pap, A., 'The Concept of Absolute Emergence', *British Journal for the Philosophy of Science* **2** (1951–1952) 320–41.
Pap, A., 'Disposition Concepts and Extensional Logic', in *Minnesota Studies in the Philosophy of Science*, vol. II (ed. by H. Feigl, M. Scriven and G. Maxwell), Minneapolis, 1958, pp. 196–224.
Pap, A., *Elements of Analytic Philosophy*, New York 1949.
Pap, A., *An Introduction to the Philosophy of Science*, Glencoe, Ill. 1962.
Pap, A., 'Logic and the Concept of Entailment', *Journal of Philosophy* **47** (1950) 378–87.
Pap, A., 'Logic and the Synthetic A Priori', *Philosophy and Phenomenological Research* **10** (1949–1950) 500–14.
Pap, A., 'Logical Nonsense', *Philosophy and Phenomenological Research* **9** (1948–1949) 269–83.
Pap, A., 'Mathematics, Abstract Entities and Modern Semantics', *Scientific Monthly* **85** (1957) 29–40.
Pap, A., 'Once More: Colors and the Synthetic *A Priori*', *Philosophical Review* **66** (1957) 94–9.
Pap, A., 'Philosophical Analysis, Translation Schemes and the Regularity Theory of Causation', *Journal of Philosophy* **49** (1952) 657–66.
Pap, A., 'Reduction Sentences and Disposition Concepts', *The Philosophy of Rudolf Carnap* (ed. by P. A. Schilpp), La Salle, Ind., 1963, pp. 559–98.
Pap, A., *Semantics and Necessary Truth*, New Haven, Conn. 1958.
Pap, A., 'Synonymy, Identity of Concepts and the Paradox of Analysis', *Methodos* **7** (1955) 115–28.
Pap, A., 'The Concept of Absolute Emergence', *British Journal for the Philosophy of Science* **2** (1951–52) 302–11.
Pap, A., 'Theory of Definition', *Philosophy of Science* **31** (1964) 49–54.
Pap, A., 'Types and Meaninglessness', *Journal of Philosophy* **54** (1957) 778–9.
Parsons, K. P., 'On Criteria for Meaning Change', *British Journal for the Philosophy of Science* **22** (1971) 131–44.
Parsons, T., 'Various Extensional Notions of Ontological Commitment', *Philosophical Studies* **21** (1970) 65–74.
Pasch, A., 'Empiricism: One "Dogma" or Two?' *Journal of Philosophy* **53** (1956) 302–11.
Patin, H. A., 'Pragmatism, Intuitionism, and Formalism', *Philosophy of Science* **24** (1957) 243–52.
Pauli, W. (ed.), *Neils Bohr and the Development of Physics*, New York 1955.
Peach, B., 'A Nondescriptive Theory of the Analytic', *Philosophical Review* **61** (1952) 349–367.
Peddie, W., 'The Philosophy of "As If" in Physical Science', *Philosophy of Science* **6** (1939) 38–47.
Peirce, C. S., 'What is a Leading Principle?', *The Philosophy of Peirce* (ed. by J. Buchler), London, 1940, pp. 129–134.
Pepper, S. C., 'Categories', *California University Publications in Philosophy* **13** (1930) 73–98.

Pepper, S. C., 'On the Cognitive Value of World Hypotheses', *Journal of Philosophy* **33** (1936) 575–7.
Pepper, S. C., 'What are Categories for?', *Journal of Philosophy* **44** (1947) 546–56.
Pepper, S. C., *World Hypotheses*, Berkeley, Calif. 1942.
Perkins, M. and Singer, I., 'Analyticity', *Journal of Philosophy* **48** (1951) 485–97.
Piaget, J., *The Child's Conception of Geometry*, London 1960.
Piaget, J., *The Child's Conception of Number*, London 1952.
Piaget, J., *The Child's Conception of Physical Causality*, 4th ed., London 1965.
Piaget, J., *The Child's Conception of the World*, London 1929.
Piaget, J., *Genetic Epistemology*, New York 1970.
Piaget, J., *Judgement and Reasoning in the Child*, London 1928.
Piaget, J., *Language and Thought of the Child*, London 1926.
Piaget, J., *Logic and Psychology*, Manchester, England 1953.
Piaget, J., *Mechanisms of Perception*, New York 1969.
Piaget, J., *The Moral Judgment of the Child*, London 1950.
Piaget, J., *The Origin of Intelligence in the Child*, London 1953.
Piaget, J., *Psychology of Intelligence*, London 1950.
Piaget, J., *Structuralisme*, Paris 1968.
Piaget, J., *Traité de logique*; *Essai de logistique aperataire*, Paris 1949.
Pirie, N. W., 'Concepts out of Context', *British Journal for the Philosophy of Science* **2** (1951–1952) 269–81.
Plaut, H. C., 'Empiricism, Solipsism, and Realism', *British Journal for the Philosophy of Science* **13** (1962–1963) 216–28.
Poincaré, H., *The Foundations of Science*, New York 1921.
Poincaré, H., *Science and Hypothesis*, New York 1952.
Polanyi, M., 'Science and Reality', *British Journal for the Philosophy of Science* **18** (1967–1968) 177–97.
Pollock, J. L., 'Criteria and Our Knowledge of the Material World', *Philosophical Review* **76** (1967) 28–60.
Popper, K. R., *Conjectures and Refutations*, New York 1962.
Popper, K. R., 'Indeterminism in Quantum Physics and in Classical Physics I and II', *British Journal for the Philosophy of Science* **1** (1950–1951) 117–33, 173–95.
Popper, K. R., *The Logic of Scientific Discovery*, London 1959.
Popper, K. R., *On the Sources of Knowledge and of Ignorance*, London 1960.
Post, H. R., 'Simplicity in Scientific Theories', *British Journal for the Philosophy of Science* **11** (1960–1961) 32–41.
Postal, P. M., *Constituent Structure: A Study of Contemporary Models of Syntactic Description*, 2nd ed., Bloomington, Ind. 1967.
Pour-El, M. B., 'Effectively Extensible Theories', *Journal of Symbolic Logic* **33** (1968) 56–68.
Price, H. H., *Thinking and Experience*, London 1953.
Price, H. H., *Thinking and Representation*, London 1946.
Price, R. A., 'A Note on Likeness of Meaning', *Analysis* **11** (1950–1951) 18–9.
Prior, A. N., 'Possible Worlds', *Philosophical Quarterly* **12** (1962) 36–43.
Puhvel, J. (ed.), *Substance and Structure of Language*, Berkeley, Calif. 1969.
Pumphrey, R. K., 'The Evolution of Thinking', *British Journal for the Philosophy of Science* **4** (1953–1954) 315–27.
Purtill, R. L., 'Kuhn on Scientific Revolutions', *Philosophy of Science* **34** (1967) 53–8.

Putnam, H., 'The Analytic and the Synthetic', in *Minnesota Studies in the Philosophy of Science*, vol. III (ed. by H. Feigl and G. Maxwell) Minneapolis, 1962, pp. 358–97.
Putnam, H., 'A Definition of Degree of Confirmation for Very Rich Languages', *Philosophy of Science* 23 (1956) 58–62.
Putnam, H., 'Dreaming and Depth Grammer', in *Analytical Philosophy* (ed. by R. J. Bulter), Oxford, 1962, pp. 211–35.
Putnam, H., 'An Examination of Grünbaum's Philosophy of Geometry', *Delaware Seminar in the Philosophy of Science*, vol. 2, New York, 1963, pp. 205–55.
Putnam, H., 'Formalization of the Concept "About"', *Philosophy of Science* 25 (1958) 125–30.
Putnam, H., 'The "Innateness Hypothesis" and Explanatory Models in Linguistics', *Synthese* 17 (1967) 12–22.
Putnam, H., 'Is Logic Empirical?', in *Boston Studies in the Philosophy of Science*, vol. V (ed. by R. Cohen and M. Wartofsky) Dordrecht, 1969, pp. 216–41.
Putnam, H., 'It Ain't Necessarily So', *Journal of Philosophy* 59 (1962) 658–71.
Putnam, H., 'Mathematics Without Foundations', *Journal of Philosophy* 64 (1967) 5–22.
Putnam, H., 'Minds and Machines', in *Dimensions of Mind* (ed. by S. Hook), New York, 1960, pp. 138–64.
Putnam, H., 'On Hierarchies and Systems of Notations', *American Mathematical Society, Proceedings* 15 (1964) 44–50.
Putnam, H., 'The Mental Life of Some Machines', in *Intentionality, Minds and Perception* (ed. by H. Castañeda) Detroit, 1966, pp. 177–200.
Putnam, H., 'Red, and Green All Over Again: A Rejoinder to Arthur Pap', *Philosophical Review* 66 (1957) 100–6.
Putnam, H., 'Reds, Greens and Logical Analysis', *Philosophical Review* 65 (1956) 206–17.
Putnam, H., 'Robots: Machines or Artificially Created Life?', *Journal of Philosophy* 61 (1964) 668–91.
Putnam, H., 'Synonymity and the Analysis of Belief Sentences', *Analysis* 14 (1953–1954) 114–23.
Putnam, H., 'Time and Physical Geometry', *Journal of Philosophy* 64 (1967) 240–7.
Putnam, H. and Ullian, J., 'More about "About"', *Journal of Philosophy* 62 (1965) 305–10.
Quine, W. V., 'On the Reasons for Indeterminacy of Translation', *The Journal of Philosophy* 67 (1970) 179–81.
Quine, W. V., 'Philosophical Progress in Language Theory', in *Language, Belief and Metaphysics* (ed. by H. Keifer and M. Munitz), Albany 1970, pp. 3–18.
Quine, W. V., *Set Theory and its Logic*, Cambridge, Mass. 1963.
Quine, W. V. O., 'Carnap and Logical Truth', *Logic and Language, Studies Dedicated to Professor Rudolf Carnap*, Dordrecht, 1962, pp. 39–63.
Quine, W. V. O., 'Designation and Existence', *Journal of Philosophy* 36 (1939) 701–9.
Quine, W. V. O., *From a Logical Point of View*, Cambridge, Mass. 1961.
Quine, W. V. O., 'Meaning and Translation', *On Translation* (ed. by R. A. Brouwer), Cambridge, Mass., 1959, pp. 148–72.
Quine, W. V. O., *Methods of Logic*, New York 1950.
Quine, W. V. O., 'Natural Kinds', *Essays in Honor of Carl G. Hempel* (ed. by N. Rescher) Dordrecht, 1970, pp. 5–23.
Quine, W. V. O., 'Necessary Truth', *Philosophy of Science Today* (ed. by S. Morgenbesser), New York, 1967, pp. 46–54.

Quine, W. V. O., 'Notes on Existence and Necessity', *Journal of Symbolic Logic* **10** (1945) 1–12.
Ouine, W. V. O., 'On a Suggestion of Katz', *Journal of Philosophy* **64** (1967) 52–4.
Quine, W. V. O., *Ontological Relativity and Other Essays*, New York 1969.
Quine, W. V. O., 'The Scope and Language of Science', *British Journal for the Philosophy of Science* **8** (1957–1958) 1–17.
Quine, W. V. O., 'Truth by Convention', *Readings in Philosophical Analysis* (ed. by H. Feigl and W. Sellars), New York, 1949, pp. 250–76.
Quine, W. V. O., *The Ways of Paradox and Other Essays*, New York 1966.
Quine, W. V. O., *Word and Object*, Cambridge, Mass. 1960.
Rasiowa, H., 'Constructive Theories', *Academie Polonaise des Sciences, Bulletin*, Classe III, Bd. 2 (1954) 121–24.
Rasiowa, H. and Sikorski, R., *The Mathematics of Metamathematics*, Warsaw 1963.
Ratner, J., 'Scientific Objects and Empirical Things', *Journal of Philosophy* **32** (1935) 393–408.
Reichenbach, H., *Philosophic Foundations of Quantum Mechanics*, Berkeley, Calif. 1948.
Reichenbach, H., 'The Philosophical Significance of the Theory of Relativity', in *Readings in the Philosophy of Science* (ed. by H. Feigl and M. Brodbeck), New York, 1953, pp. 195–211.
Reichenbach, H., *The Philosophy of Space and Time*, (transl. by M. Reichenbach and J. Freund), New York 1958.
Reichenbach, H., *The Rise of Scientific Philosophy*, Berkeley, Calif. 1951.
Reiser, O. L., 'The Evolution of Cosmologies', *Philosophy of Science* **19** (1952) 93–107.
Rescher, N., 'The Identity of Indiscernibles', *Journal of Philosophy* **52** (1955) 152–5.
Rescher, N., 'A New Look at the Problem of Innate Ideas', *British Journal for the Philosophy of Science* **17** (1966–1967) 205–19.
Rescher, N., 'Translation as a Tool of Philosophical Analysis', *Journal of Philosophy* **53** (1956) 219–24.
Rescher, N. (ed.), *Essays in Honor of Carl G. Hempel*, Dordrecht 1970.
Rescher, N. (ed.), *Studies in the Philosophy of Science*, Oxford 1969.
Resnik, L., 'Empiricism and Natural Kinds', *Journal of Philosophy* **57** (1960) 555–9.
Robinett, F. M., Voegelin, C. F., and Yegerlehner, J. F., 'Shawnee Laws: Perceptual Statements for the Language and for the Content', *Language in Culture* (ed. by H. Hoijer), Chicago, Ill. 1954, pp. 32–46.
Robbins, B., 'Ontology and the Hierarchy of Languages', *Philosophical Review* **67** (1958) 531–7.
Rogers, R., 'Mathematical and Philosophical Analysis', *Philosophy of Science* **31** (1964) 255–64.
Rollins, C. D., 'The Philosophical Denial of Sameness of Meaning', *Analysis* **11** (1950–1951) 38–45.
Rorty, R., 'Mind-Body Identity, Privacy and Categories', *Review of Metaphysics* **19** (1965–1966) 24–54.
Rorty, R. M., 'Pragmatism, Categories and Language', *Philosophical Review* **70** (1961) 197–223.
Rorty, R. M., 'Realism, Categories and the "Linguistic Turn"', *International Philosophical Quarterly* **2** (1962) 307–22.
Rorty, R. M. (ed.), *The Linguistic Turn*, Chicago 1967.
Rosenberg, J. F., 'Synonymy and the Epistemology of Linguistics', *Inquiry* **4** (1967) 405–20.

Rosenblueth, A. and Wiener, N., 'The Role of Models in Science', *Philosophy of Science* **12** (1945) 316–21.
Ross, J. F., 'Logically Necessary Existential Statements', *Journal of Philosophy* **58** (1961) 253–63.
Rougier, L., 'The Relativity of Logic', *Philosophy and Phenomenologocal Research* **2** (1941–1942) 137–57.
Rozeboom. W. W., 'The Factual Content of Theoretical Concepts', in *Minnesota Studies in the Philosophy of Science*, vol. III (ed. by H. Feigl and G. Maxwell), Minneapolis, 1962, pp. 273–357.
Rozeboom, W. W., 'Of Selection Operators and Semanticists', *Philosophy of Science* **31** (1964) 282–85.
Rozeboom, W. W., 'Ontological Induction and the Logical Topology of Scientific Variables', *Philosophy of Science* **28** (1961) 337–77.
Rudner, R. S., 'Formal and Non-Formal', *Philosophy of Science* **16** (1949) 41–8.
Rudner, R. S., 'An Introduction to Simplicity', *Philosophy of Science* **28** (1961) 109–19.
Rudner, R. S., 'A Note on Likeness of Meaning', *Analysis* **10** (1949–1950) 115–8.
Russell, B., *An Inquiry into Meaning and Truth*, London 1940.
Ryle, G., 'Categories', *Aristotelian Society, Proceedings* **38** (1937–1938) 186–206.
Ryle, G., *The Concept of Mind*, New York 1949.
Ryle, G., *Dilemmas*, Cambridge, England 1954.
Ryle, G., 'Ordinary Language', *Philosophical Review* **62** (1953) 167–86.
Ryle, G., 'Philosophical Arguments', in *Logical Positivism* (ed. by A. J. Ayer), New York, 1959, pp. 327–44.
Ryle, G., 'Systematically Misleading Expressions', *Aristotelian Society, Proceedings* **32** (1931–1932) 139–70.
Ryle, G., 'Unverifiability by Me', *Analysis* **4** (1936–1937) 1–11.
Sanford, D., 'Disjunctive Predicates', *American Philosophical Quarterly* **7** (1970) 162–70.
Scheffler, I., *The Anatomy of Inquiry: Philosophical Studies in the Theory of Science*, New York 1963.
Scheffler, I., *Conditions of Knowledge*, Chicago 1965.
Scheffler, I., 'Prospects of a Modest Empiricism', *Review of Metaphysics* **10** (1957) 383–400, 602–25.
Scheffler, I., *Science and Subjectivity*, Indianapolis 1967.
Scheffler, I., 'Supercalifragilistic Reduction, A Reply to Jan Berg', *Philosophy of Science* **38** (1971) 121–2.
Schilpp, P. A., 'The Nature of the "Given"', *Philosophy of Science* **2** (1935) 128–38.
Schlesinger, G., 'The Formalization of Empirical Significance', *Philosophy of Science* **31** (1964) 65–7.
Schlesinger, G., 'What Does the Denial of Absolute Space Mean?', *Australasian Journal of Philosophy* **45** (1967) 44–60.
Schlick, M., 'Causality in Contemporary Physics', *British Journal for the Philosophy of Science* **12** (1961–1962) 177–93, 281–99.
Schlick, M., 'Is There a Factual A Priori?', in *Readings in Philosophical Analysis* (ed. by H. Feigl and W. Sellars), New York, 1949, pp. 277–85.
Schuldenfrei, R., 'Quine in Perspective', *Journal of Philosophy* **69** (1972) 5–16.
Scott, D. and Suppes, P., 'Foundational Aspects of Theories of Measurement', *Journal of Symbolic Logic* **23** (1958) 113–28.

Scriven, M. J., 'Definitions, Explanations, and Theories', in *Minnesota Studies in the Philosophy of Science*, Vol. II (ed. by H. Feigl, M. Scriven, and G. Maxwell), Minneapolis, 1958, pp. 99-195.
Scriven, M. J., 'Explanation and Predication in Evolutionary Theory', *Science* **130** (1959) 477-82.
Scriven, M. J., 'A Possible Distinction between Traditional Scientific Disciplines and the Study of Human Behavior', in *Minnesota Studies in the Philosophy of Science*, vol. I (ed. by H. Feigl and M. Scriven), Minneapolis, 1956, pp. 330-9.
Sellars, W., 'Abstract Entities', *Review of Metaphysics* **19** (1963) 627-71.
Sellars, W., 'Concepts as Involving Laws and Inconceivable without Them', *Philosophy of Science* **15** (1948) 287-313.
Sellars, W., 'Counterfactuals, Dispositions and the Causal Modalities', in *Minnesota Studies in the Philosophy of Science*, vol. II (ed. by H. Feigl, M. Scriven, and G. Maxwell), Minneapolis, 1958, pp. 225-308.
Sellars, W., 'Empiricism and the Philosophy of Mind', in *Minnesota Studies in the Philosophy of Science*, Vol. I (ed. by H. Feigl, and M. Scriven), Minneapolis, 1956, pp. 253-329.
Sellars, W., 'Induction as Vindication', *Philosophy of Science* **31** (1964) 197-231.
Sellars, W., 'Inference and Meaning', *Mind* **62** (1953) 313-38.
Sellars, W., 'Is There a Synthetic A Priori?', *Philosophy of Science* **20** (1953) 121-38.
Sellars, W., 'Language as Thought and as Communication', *Philosophy and Phenomenological Research* **30** (1969) 506-27.
Sellars, W., 'Philosophy and the Scientific Image of Man', *Frontiers of Science and Philosophy* (ed. by R. G. Colodny), Pittsburgh, 1962, pp. 35-78.
Sellars, W., *Philosophical Perspectives*, Springfield 1959.
Sellars, W., 'Realism and the New Way of Words', *Philosophy and Phenomenological Research* **8** (1947-1948) 601-34.
Sellars, W., *Science and Metaphysics: Variations on Kantian Themes*, New York 1968.
Sellars, W., *Science, Perception, and Reality*, London 1963.
Sellars, W., 'Science, Sense Impressions and Sense: A Reply to Cornman', *Review of Metaphysics* **24** (1970) 391-447.
Sellars, W., 'Scientific Realism and Irenic Instrumentalism', in *Boston Studies in the Philosophy of Science*, vol. II (ed. by R. S. Cohen and M. W. Wartofsky), New York, (1965) pp. 171-204.
Sellars, W., 'Some Reflections on Language Games', *Philosophy of Science* **21** (1954) 204-28.
Sellars, W., 'Some Remarks on Kant's Theory of Experience', *Journal of Philosophy* **64** (1967) 633-47.
Sellars, W., 'Time and the World Order', in *Minnesota Studies in the Philosophy of Science*, vol. III (ed. by H. Feigl and G. Maxwell), Minneapolis, 1962, pp. 527-616.
Sellars, W., 'Towards a Metaphysics of the Person', *The Logical Way of Doing Things* (ed. by K. Lambert), New Haven, 1969, pp. 219-52.
Sellars, R. W., 'Verification of Categories: Existence and Substance', *Journal of Philosophy* **40** (1943) 197-205.
Sellars, W. and Feigl, H. (eds.), *Readings in Philosophical Analysis*, New York 1949.
Settle, T. W., 'The Point of Positive Evidence – Reply to Professor Feyerabend', *British Journal for the Philosophy of Science* **20** (1969) 352-4.
Shapere, D., 'Meaning and Scientific Change', in *Mind and Cosmos* (ed. by R. Colodny) Pittsburgh, 1966, pp. 41-85.

Shapere, D., 'Plausibility and Justification in the Development of Science', *Journal of Philosophy* **63** (1966) 611–21.
Shapere, D., 'Space, Time and Language – An Examination of Some Problems and Methods of the Philosophy of Science', *Delaware Seminar in the Philosophy of Science*, vol. **2**, New York, 1963, pp. 139–70.
Shapere, D., 'The Structure of Scientific Revolutions', *Philosophical Review* **73** (1964) 383–94.
Shimony, A., 'Scientific Inference', *The Nature and Function of Scientific Theories* (ed. by R. G. Colodny), Pittsburgh, 1970, pp. 79–172.
Sibley, C., 'The Problem of the Categories', *Philosophical Review* **45** (1936) 283–96.
Sibley, W. M., 'The Pragmatic Theory of Scientific Objects', *Philosophical Review* **58** (1948) 248–59.
Sklar, L., 'The Falsifiability of Geometric Theories', *Journal of Philosophy* **64** (1967) 247–53.
Sklar, L., 'Types of Inter-Theoretic Reduction', *British Journal for the Philosophy of Science* **18** (1967–1968) 109–24.
Sluckin, W. and Thompson, R., 'Cybernetics and Mental Functioning', *British Journal for the Philosophy of Science* **4** (1953–1954) 130–46.
Smart, H. R., 'Language-Games', *Philosophical Quarterly* **7** (1957) 224–35.
Smart, J. J. C., 'Sensations and Brain Processes', in *The Philosophy of Mind* (ed. by V. C. Chappell), Englewood Cliffs, N.J., 1962, pp. 160–72.
Smart, J. J. C., 'Theory Construction', *Philosophy and Phenomenological Research* **11** (1950–1951) 457–73.
Smith, M., 'Hypothesis vs. Problem in Scientific Investigation', *Philosophy of Science* **12** (1945) 296–301.
mullyan, A., 'The Concept of Empirical Knowledge', *Philosophical Review* **65** (1956) 362–70.
Sommers, F., 'Meaning Relations and the Analytic', *Journal of Philosophy* **60** (1963) 524–34.
Sparkes, J. J., 'Pattern Recognition and Scientific Progress', *Mind* **81** (1972) 29–74.
Spector, M., 'Models and Theories', *British Journal for the Philosophy of Science* **16** (1965) 121–42.
Spector, M., 'Theory and Observation', *British Journal for the Philosophy of Science* **17** (1965–1966) 1–20, 89–104,
Stadler, I. H., 'On "Seeing As"', *Philosophical Review* **67** (1958) 91–4.
Stahl, G., 'Linguistic Structures Isomorphic to Object Structures', *Philosophy and Phenomenological Research* **24** (1963–1964) 339–43.
Stallo, J. B., *Concepts of Theories of Modern Physics*, Cambridge, Mass. 1960.
Stannard, J., 'The Role of Categories in Historical Explanation', *Journal of Philosophy* **56** (1959) 429–47.
Stoppes-Roe, H. V., 'Some Considerations Concerning "Interpretive Systems"', *Philosophy of Science* **25** (1958) 143–56.
Storer, T., 'Linguistic Isomorphisms', *Philosophy of Science* **19** (1952) 77–85.
Strauss, M., 'Einstein's Theories and the Critics of Newton: An Essay in Logico-Historical Analysis', *Synthese* **18** (1968) 251–84.
Strauss, M., 'Intertheory Relations', *Proceedings of the 1968 Salzburg Colloquium in Philosophy of Science*, Dordrecht 1970, pp. 220–84.
Strawson, P. F., *The Bounds of Sense: An Essay on Kant's 'Critique of Pure Reason'*, London 1966.

Strawson, P. F., *Individuals: An Essay in Descriptive Metaphysics*, London 1959.
Strawson, P. F., 'Logical Subjects and Physical Objects, Symposium', *Philosophy and Phenomenological Research* **17** (1956–1957) 441–57.
Strawson, P. F., 'Necessary Propositions and Entailment-Statements', *Mind* **57** (1948) 184–200.
Strawson, P. F., 'Paradoxes, Posits and Propositions', *Philosophical Review* **76** (1967) 214–9.
Strawson, P. F., 'Persons', in *Minnesota Studies in the Philosophy of Science*, vol. II (ed. by H. Feigl, M. Scriven, and G. Maxwell), Minneapolis, 1958, pp. 330–52.
Stroud, B., 'Conventionalism and the Indeterminacy of Translation', *Synthese* **19** (1969) 82–96.
Stroud, B., 'Wittgenstein and Logical Necessity', *Philosophical Review* **74** (1965) 504–18.
Suppes, P., 'The Desirability of Formalization in Science', *Journal of Philosophy* **65** (1968) 651–64.
Suppes, P., *Studies in the Methodology and Foundations of Science: Selected Papers from 1951 to 1969*, New York 1969.
Suppes, P., 'What is a Scientific Theory?', in *Philosophy of Science Today* (ed. by S. Morgenbesser), New York, 1967, pp. 55–67.
Swanson, J. W., 'Linguistic Relativity and Translation', *Philosophy and Phenomenological Research* **22** (1961–1962) 185–92.
Swanson, J. W., 'On Models', *British Journal for the Philosophy of Science* **17** (1966–1967) 297–312.
Swinburne, R. G., 'Choosing between Confirmation Theories', *Philosophy of Science* **37** (1970) 602–13.
Swinburne, R. G., 'Popper's Account of Acceptability', *Australasian Journal of Philosophy* **49** (1971) 167–76.
Tabory, R., 'Semantics, Generative Grammars, and Computers. A Few Comments on Current Trends in Semantics', *Linguistics* **16** (1965) 68–85.
Tanner, R. C. H., 'On the Role of Equality and Inequality in the History of Mathematics', *British Journal for the History of Science* **1** (1962–1963) 159–70.
Taylor, R., 'Disputes about Synonymy', *Philosophical Review* **63** (1954) 517–29.
Taylor, R., 'Spatial and Temporal Analogies and the Concept of Identity', *Journal of Philosophy* **52** (1955) 599–612.
Tennessen, H., 'On Making Sense', *Journal of Philosophy* **57** (1960) 764–5.
Tennessen, H., 'Permissible and Impermissible Locutions', in *Logic and Language, Studies Dedicated to Professor Rudolf Carnap*, Dordrecht, 1962, pp. 220–33.
Tennessen, H., 'What Should We Say?', *Inquiry* **11** (1968) 228.
Thomason, J. M., 'Ontological Relativity and the Inscrutability of Reference', *Philosophical Studies* **22** (1971) 50–6.
Thompson, M., 'On Category Differences', *Philosophical Review* **66** (1957) 486–508.
Tillman, F., 'Linguistic Portrayal and Theoretical Involvement', *Philosophy and Phenomenological Research* **27** (1966–1967) 597–605.
Tisza, L., 'The Conceptual Structure of Physics', *Reviews of Modern Physics* **35** (1963) 151–85.
Toulmin, S. E., 'Conceptual Revolutions in Science', *Synthese* **17** (1967) 75–91.
Toulmin, S. E., *The Ancestry of Science*, London 1961.
Toulmin, S. E., *Discovery of Time*, New York 1965.
Toulmin, S. E., *The Fabric of the Heavens: The Development of Astronomy and Dynamics*, New York 1962.

GENERAL BIBLIOGRAPHY

Toulmin, S. E., *Metaphysical Beliefs; Three Essays* (ed. by A. MacIntyre), London, 1957.
Turner, J., 'Maxwell on the Logic of Dynamical Explanation', *Philosophy of Science* **23** (1956) 36–47.
Turquette, A. R., 'Gödel and the Synthetic A Priori', *Journal of Philosophy* **47** (1950) 125–9.
Ullian, J. S., 'Is Any Set Theory True?', *Philosophy of Science* **36** (1969) 271–9.
Ullian, J. S., 'Luck, Licence and Lingo', *Journal of Philosophy* **58** (1961) 731–8.
Ullian, J. S., 'Mathematical Objects', *Delaware Seminar in the Philosophy of Science*, Vol. I, New York 1963, pp. 187–205.
Ullian, J. S., 'More on "Grue" and Grue', *Philosophical Review* **70** (1961) 731–8.
Ullman, S., *The Principles of Semantics*, Glasgow 1951.
Umezawa, T., 'On Logics Intermediate between Intuitionistic and Classical Predicate Logic', *Journal of Symbolic Logic* **24** (1959) 141–53.
von Wright, G. H., *Logical Studies*, London 1957.
von Wright, G. H., *Time, Change and Contradiction*, London 1969.
Waismann, F., 'Analytic-Synthetic', *Analysis* **10** (1949) 25–40; **11** (1950) 25–38; **11** (1951) 166–24; **13** (1952) 1–14; **13** (1952) 49–61; **11** (1951) 74–80.
Waismann, F., 'Language Strata', in *Logic and Language* (ed. by A. Flew), 2nd, series Garden City, N.Y., 1965, pp. 226–47.
Waismann, F., 'Verifiability', in *Logic and Language* (ed. by A. Flew), 1st series, Garden City, N.Y., 1965, pp. 122–55.
Wallace, J., 'A Query on Radical Translation', *J. of Philosophy* **68** (1971) 143–51.
Walsh, D., 'Literature and the Categories', *Journal of Philosophy* **55** (1958) 846–55.
Walton, K. L., *Conceptual Schemes: A Study of Linguistic Relativity and Related Philosophical Problems*, Ph.D. Thesis, Cornell University, 1967.
Warnock, G. L., 'Concepts and Schematism', *Analysis* **9** (1948–1949) 77–83.
Wein, H., 'The Categories and a Logic of Structure', *Journal of Philosophy* **49** (1952) 629–33.
Weinberg, J. K., *An Examination of Logical Positivism*, London 1936.
Weinpahl, P., 'More about the Denial of Sameness of Meaning', *Analysis* **12** (1951–1952) 19–24.
Weissman, A., 'The Meaning of Identity', *Philosophy and Phenomenological Research* **16** (1955–1956) 461–75.
Wellman, C., 'Wittgenstein's Conception of a Criterion', *Philosophical Review* **71** (1962) 433–77.
Wells, D. A., 'Some Implications of Empirical Truth by Convention', *Journal of Philosophy* **48** (1951) 185–92.
Wells, R., 'What has Linguistics Done for Philosophy?', *J. of Philosophy* **59** (1962) 697–708.
Werkmeister, W. H., 'Natural Languages as Cultural Indices', *Philosophy of Science* **6** (1939) 356–66.
Werkmeister, W. H., 'Science, Its Concepts and Laws', *Journal of Philosophy* **46** (1949) 444–52.
Whewell, W., *The Philosophy of the Inductive Sciences*. vols. I and II. Facsimile of 2nd ed., London 1967.
White, M., 'The Analytic and the Synthetic: an Untenable Dualism', in *Semantics and the Philosophy of Language* (ed. by L. Linsky) Urbana, Ill., 1952, pp. 272–86.
Whitehead, A. N., *The Concept of Nature*, Cambridge, England 1920.

Whitehead, A. N., *A Philosopher Looks at Science*, New York 1965.
Whitehead, A. N., *The Principle of Relativity*, Cambridge, England 1922.
Whitehead, A. N., *Science and the Modern World*, New York 1925.
Whorf, B. L., *The Hopi Language*, Chicago, Ill. 1956.
Whorf, B. L., *Language, Thought and Reality: Selected Writings of Benjamin Lee Whorf* (ed. by J. B. Carroll), New York 1956.
Whorf, B. L., 'Language, Mind and Reality', *ETC, A Review of General Semantics* **9** (1951) 167–88.
Wick, W. A., 'Minds, Artificial Languages and Philosophy', *Philosophy and Phenomenological Research* **14** (1953–1954) 228–38.
Wiener, P. P., 'Philosophical, Scientific, and Ordinary Language', *Journal of Philosophy* **45** (1948) 260–7.
Wigner, E. P., *Symmetries and Reflections: Scientific Essays* (ed. by M. J. Scriven and W. J. Moors), Bloomington, 1967.
Wilson, J. C., 'Categories in Aristotle and Kant', in *Aristotle, a Collection of Critical Essays* (ed. by J. M. E. Moravcsik), Garden City, N.Y., 1967, pp. 75–89.
Wilson, N. L., 'Grice on Meaning: The Ultimate Counterexample', *Noûs* **4** (1970) 295–302.
Wilson, N. L., 'Linguistic Butter and Philosophical Parsnips', *Journal of Philosophy* **64** (1967) 55–67.
Wilson, N. L., 'Substances Without Substrata', *Review of Metaphysics* **12** (1959) 521–39.
Wilson, N. L., *The Concept of Language*, Toronto 1959.
Wilson, N. L., 'The Indestructibility and Immutability of Substances', *Philosophical Studies* **7** (1956) 46–8.
Wilson, N. L., 'The Trouble With Meanings', *Dialogue* **4** (1964) 52–64.
Winnie, J. A., 'The Implicit Definition of Theoretical Terms', *British Journal for the Philosophy of Science* **18** (1967–1968) 223–9.
Wisdom, J., *Logical Constructions*, New York 1969.
Wisdom, J., *Paradox and Discovery*, New York 1965.
Wisdom, J. O., 'The Hypothesis of Cybernetics', *British Journal for the Philosophy of Science* **2** (1951–1952) 1–24.
Wittgenstein, L., *The Blue and Brown Books*, Oxford 1958.
Wittgenstein, L., *Philosophical Investigations*, New York 1953.
Wittgenstein, L., *Remarks on the Foundations of Mathematics* (ed. by G. H. von Wright, R. Rhees and G. E. M. Anscombe), Cambridge, Mass. 1967.
Woodger, J. H., *The Technique of Theory Construction*, Chicago, Ill. 1939.
Woodger, J. H., 'What Do We Mean by "In Born"?', *British Journal for the Philosophy of Science* **3** (1952–1953) 319–27.
Woolhouse, R., 'Third Possibilities and the Law of the Excluded Middle', *Mind* **76** (1967) 283–5.
Zaffron, Richard, 'Identity, Subsumption, and Scientific Explanation', *Journal of Philosophy* **68** (1971) 849–60.
Ziff, P., 'About Behaviorism', in *The Philosophy of Mind* (ed. by V. C. Chappell), Englewood Cliffs, N.J., 1962, pp. 147–50.
Ziff, P., 'Feelings of Robots', in *Minds and Machines* (ed. by A. R. Anderson), Englewood Cliffs, N.J., 1964, pp. 98–103.
Ziff, P., *Semantic Analysis*, Ithaca 1960.
Ziff, P., *Understanding Understanding*, Ithaca, 1972.

SYNTHESE LIBRARY

Monographs on Epistemology, Logic, Methodology,
Philosophy of Science, Sociology of Science and of Knowledge, and on the
Mathematical Methods of Social and Behavioral Sciences

Editors:

DONALD DAVIDSON (The Rockefeller University and Princeton University)
JAAKKO HINTIKKA (Academy of Finland and Stanford University)
GABRIËL NUCHELMANS (University of Leyden)
WESLEY C. SALMON (Indiana University)

M. BUNGE, *Exact Philosophy – Problems, Tools, and Goals.* 1973, X + 214 pp.

ROBERT S. COHEN and MARX W. WARTOFSKY (eds.), *Boston Studies in the Philosophy of Science.* Volume IX: *A. A. Zinov'ev: Foundations of the Logical Theory of Scientific Knowledge (Complex Logic).* Revised and Enlarged English Edition with an Appendix by G. A. Smirnov, E. A. Sidorenka, A. M. Fedina, and L. A. Bobrova. 1973, XXII + 301 pp. (Also in paperback).

K. J. J. HINTIKKA, J. M. E. MORAVCSIK, and P. SUPPES (eds.), *Approaches to Natural Language. Proceedings of the 1970 Stanford Workshop on Grammar and Semantics.* 1973, VIII + 526 pp. (Also in paperback).

WILLARD C. HUMPHREYS, JR. (ed.), *Norwood Russell Hanson: Constellations and Conjectures.* 1973, X + 282 pp.

MARIO BUNGE, *Method, Model and Matter.* 1973, VII + 196 pp.

MARIO BUNGE, *Philosophy of Physics.* 1973, IX + 248 pp.

LADISLAV TONDL, *Boston Studies in the Philosophy of Science.* Volume X: *Scientific Procedures.* 1973, XIII + 268 pp. (Also in paperback).

SÖREN STENLUND, *Combinators, λ-Terms and Proof Theory.* 1972, 184 pp.

DONALD DAVIDSON and GILBERT HARMAN (eds.), *Semantics of Natural Language.* 1972, X + 769 pp. (Also in paperback).

MARTIN STRAUSS, *Modern Physics and Its Philosophy. Selected Papers in the Logic, History, and Philosophy of Science.* 1972, X + 297 pp.

‡ STEPHEN TOULMIN and HARRY WOOLF (eds.), *Norwood Russell Hanson: What I Do Not Believe, and Other Essays.* 1971, XII + 390 pp.

‡ ROBERT S. COHEN and MARX W. WARTOFSKY (eds.), *Boston Studies in the Philosophy of Science.* Volume VIII: *PSA 1970. In Memory of Rudolf Carnap* (ed. by Roger C. Buck and Robert S. Cohen). 1971, LXVI + 615 pp. (Also in paperback).

‡ YEHOSUA BAR-HILLEL (ed.), *Pragmatics of Natural Languages.* 1971, VII + 231 pp.

‡ ROBERT S. COHEN and MARX W. WARTOFSKY (eds.), *Boston Studies in the Philosophy*

of Science. Volume VII: *Milič Čapek: Bergson and Modern Physics.* 1971, XV + 414 pp.

‡CARL R. KORDIG, *The Justification of Scientific Change.* 1971, XIV + 119 pp.

‡JOSEPH D. SNEED, *The Logical Structure of Mathematical Physics.* 1971, XV + 311 pp.

‡JEAN-LOUIS KRIVINE, *Introduction to Axiomatic Set Theory.* 1971, VII + 98 pp.

‡RISTO HILPINEN (ed.), *Deontic Logic: Introductory and Systematic Readings.* 1971, VII + 182 pp.

‡EVERT W. BETH, *Aspects of Modern Logic.* 1970, XI + 176 pp.

‡PAUL WEINGARTNER and GERHARD ZECHA, (eds.), *Induction, Physics, and Ethics, Proceedings and Discussions of the 1968 Salzburg Colloquium in the Philosophy of Science.* 1970, X + 382 pp.

‡ROLF A. EBERLE, *Nominalistic Systems.* 1970, IX + 217 pp.

‡JAAKKO HINTIKKA and PATRICK SUPPES, *Information and Inference.* 1970, X + 336 pp.

‡KAREL LAMBERT, *Philosophical Problems in Logic. Some Recent Developments.* 1970, VII + 176 pp.

‡P. V. TAVANEC (ed.), *Problems of the Logic of Scientific Knowledge.* 1969, XII + 429 pp.

‡ROBERT S. COHEN and RAYMOND J. SEEGER (eds.), *Boston Studies in the Philosophy of Science.* Volume VI: *Ernst Mach: Physicist and Philosopher.* 1970, VIII + 295 pp.

‡MARSHALL SWAIN (ed.), *Induction, Acceptance, and Rational Belief.* 1970, VII + 232 pp.

‡NICHOLAS RESCHER et al. (eds.), *Essays in Honor of Carl G. Hempel. A Tribute on the Occasion of his Sixty-Fifth Birthday.* 1969, VII + 272 pp.

‡PATRICK SUPPES, *Studies in the Methodology and Foundations of Science. Selected Papers from 1911 to 1969,* 1969, XII + 473 pp.

‡JAAKKO HINTIKKA, *Models for Modalities. Selected Essays.* 1969, IX + 220 pp.

‡D. DAVIDSON and J. HINTIKKA (eds.), *Words and Objections: Essays on the Work of W. V. Quine.* 1969, VIII + 366 pp..

‡J. W. DAVIS, D. J. HOCKNEY and W. K. WILSON (eds.), *Philosophical Logic.* 1969, VIII + 277 pp.

‡ROBERT S. COHEN and MARX W. WARTOFSKY (eds.), *Boston Studies in the Philosophy of Science,* Volume V: *Proceedings of the Boston Colloquium for the Philosophy of Science 1966/1968,* VIII + 482 pp.

‡ROBERT S. COHEN and MARX W. WARTOFSKY (eds.), *Boston Studies in the Philosophy of Science.* Volume IV: *Proceedings of the Boston Colloquium for the Philosophy of Science 1966/1968.* 1969, VIII + 537 pp.

‡NICHOLAS RESCHER, *Topics in Philosophical Logic.* 1968, XIV + 347 pp.

‡GÜNTHER PATZIG, *Aristotle's Theory of the Syllogism. A Logical-Philological Study of Book A of the Prior Analytics.* 1968, XVII + 215 pp.

‡ C. D. BROAD, *Induction, Probability, and Causation. Selected Papers.* 1968, XI + 296 pp.

‡ ROBERT S. COHEN and MARX W. WARTOFSKY (eds.), *Boston Studies in the Philosophy of Science.* Volume III: *Proceedings of the Boston Colloquium for the Philosophy of Science 1964/1966.* 1967, XLIX + 489 pp.

‡ GUIDO KÜNG, *Ontology and the Logistic Analysis of Language. An Enquiry into the Contemporary Views on Universals.* 1967, XI + 210 pp.

*EVERT W. BETH and JEAN PIAGET, *Mathematical Epistemology and Psychology.* 1966, XXII + 326 pp.

*EVERT W. BETH, *Mathematical Thought. An Introduction to the Philosophy of Mathematics.* 1965, XII + 208 pp.

‡ PAUL LORENZEN, *Formal Logic.* 1965, VIII + 123 pp.

‡ GEORGES GURVITCH, *The Spectrum of Social Time.* 1964, XXVI + 152 pp.

‡ A. A. ZINOV'EV, *Philosophical Problems of Many-Valued Logic.* 1963, XIV + 155 pp.

‡ MARX W. WARTOFSKY (ed.), *Boston Studies in the Philosophy of Science.* Volume I: *Proceedings of the Boston Colloquium for the Philosophy of Science, 1961–1962.* 1963, VIII + 212 pp.

‡ B. H. KAZEMIER and D. VUYSJE (eds.), *Logic and Language. Studies dedicated to Professor Rudolf Carnap on the Occasion of his Seventieth Birthday.* 1962, VI + 256 pp.

*EVERT W. BETH, *Formal Methods. An Introduction to Symbolic Logic and to the Study of Effective Operations in Arithmetic and Logic.* 1962, XIV + 170 pp.

*HANS FREUDENTHAL (ed.), *The Concept and the Role of the Model in Mathematics and Natural and Social Sciences. Proceedings of a Colloquium held at Utrecht, The Netherlands, January 1960.* 1961, VI + 194 pp.

‡ P. L. GUIRAUD, *Problèmes et méthodes de la statistique linguistique.* 1960, VI + 146 pp.

*J. M. BOCHEŃSKI, *A Precis of Mathematical Logic.* 1959, X + 100 pp.

SYNTHESE HISTORICAL LIBRARY

Texts and Studies
in the History of Logic and Philosophy

Editors:

N. KRETZMANN (Cornell University)
G. NUCHELMANS (University of Leyden)
L. M. DE RIJK (University of Leyden)

LEWIS WHITE BECK (ed.), *Proceedings of the Third International Kant Congress.* 1972, XI + 718 pp.

‡KARL WOLF and PAUL WEINGARTNER (eds.), *Ernst Mally: Logische Schriften.* 1971, X + 340 pp.

‡LEROY E. LOEMKER (ed.), *Gottfried Wilhelm Leibnitz: Philosophical Papers and Letters.* A Selection Translated and Edited, with an Introduction. 1969, XII + 736 pp.

‡M. T. BEONIO-BROCCHIERI FUMAGALLI, **The Logic of Abelard.** Translated from the Italian. 1969, IX + 101 pp.

Sole Distributors in the U.S.A. and Canada:
*GORDON & BREACH, INC., 440 Park Avenue South, New York, N.Y. 10016
‡HUMANITIES PRESS, INC., 303 Park Avenue South, New York, N.Y. 10010